Primer for the Inspection and Strength Evaluation of Suspension Bridge Cables

Publication No. FHWA-IF-11-045

May 2012

Notice

This document is disseminated under the sponsorship of the U.S. Department of Transportation in the interest of information exchange. The U.S. Government assumes no liability for use of the information contained in this document. This report does not constitute a standard, specification, or regulation.

Quality Assurance Statement

The Federal Highway Administration provides high-quality information to serve Government, industry, and the public in a manner that promotes public understanding. Standards and policies are used to ensure and maximize the quality, objectivity, utility, and integrity of its information. FHWA periodically reviews quality issues and adjusts its programs and processes to ensure continuous quality improvement.

Published by Books Express Publishing
Copyright © Books Express, 2012
ISBN 978-1-78266-148-1

Books Express publications are available from all good retail and online booksellers. For publishing proposals and direct ordering please contact us at: info@books-express.com

Primer for the Inspection and Strength Evaluation of Suspension Bridge Cables

Table of Contents

FOREWORD 1
ACKNOWLEDGEMENTS 2
1.0   Introduction 3
    1.1   Documents 3
    1.2   Primer Organization 3
    1.3   Suspension Bridge Cables 4
        1.3.1   Bridge Cable Components 5
        1.3.2   Bridge Cable Protection 9
        1.3.3   Causes of Cable Deterioration 10
        1.3.4   Cable Wire Corrosion 11
2.0   Inspection Guidelines and Laboratory Testing Methods 14
    2.1   Cable Inspection 14
        2.1.1   Levels of Inspection and Inspection Intervals 15
                2.1.1.1   Inspections by Maintenance Personnel 15
                2.1.1.2   Biennial Inspections 15
                2.1.1.3   Internal Inspections 21
        2.1.2   Outline of Internal Inspections 29
        2.1.3   Outline of Inspection and Sampling 30
        2.1.4   Splicing of New Wires into the Cable 31
    2.2   Cable Inspection 32
        2.2.1   Tests of Wire Properties 32
                2.2.1.1   Specimen Preparation 32
                2.2.1.2   Tensile Tests 33
                2.2.1.3   Obtaining Data for Stress vs. Strain Curves 33
                2.2.1.4   Fractographic Examination of Suspect Wires 33

| | | 2.2.1.5 | Examination of Fracture Surface for Pre-existing Cracks | 34 |

- 2.2.2 Zinc Coating Tests 34
    - 2.2.2.1 Weight of Zinc Coating 35
    - 2.2.2.2 Preece Test for Uniformity 35
- 2.2.3 Chemical Analysis 35
- 2.2.4 Corrosion Analysis 36

3.0 Evaluation of Field and Laboratory Data overview 37
- 3.1 Mapping and Estimating Wire Deterioration 37
    - 3.1.1 Number of Rings in the Cable 37
    - 3.1.2 Number of Wires in Each Ring 38
    - 3.1.3 Wire Deterioration/Corrosion Mapping 38
    - 3.1.4 Fraction of Cable in Each Corrosion Stage 40
    - 3.1.5 Number of Broken Wires 41
- 3.2 Wire Properties 42
    - 3.2.1 Cracked Wires as a Separate Group 42
    - 3.2.2 Individual Wire Properties 43
        - 3.2.2.1 Mean Properties 43
        - 3.2.2.2 Minimum Properties in a Panel Length 44
    - 3.2.3 Wire Group Mean Strength and Standard Deviation 45
- 3.3 Wire Redevelopment 45

4.0 Estimation of Cable Strength 46
- 4.1 Wire Groupings 47
- 4.2 Strength of Unbroken Wires 47
    - 4.2.1 Simplified Strength Model 47
        - 4.2.1.1 Mean Tensile Strength of Uncracked Wires 48
        - 4.2.1.2 Cable Strength Using the Simplified Model 48
    - 4.2.2 Brittle-Wire Strength Model 49
    - 4.2.3 Limited Ductility Strength Model 50
- 4.3 Non-applicability of Load and Resistance Factor Rating (LRFR) 51

5.0 Inspection Documentation, Reporting, and Recommendations 52
- 5.1 Maintenance Personnel Inspection Reports 52

5.2	Biennial Inspection Report	52
5.3	Internal Inspection Report	53
6.0	References	55
7.0	Appendix A: Strength Evaluation Example	56
7.1	Prior to Inspection	57
7.1.1	Number of Wires in the Cable	57
7.1.2	Number of Rings in the Cable	57
7.1.3	Number of Wires in Each Ring	59
7.2	Analyzing Inspection Data	60
7.2.1	Corrosion Map	63
7.2.2	Broken and Removed Wires for Testing	64
7.2.3	Number of Wires in Each Corrosion Stage	65
7.3	Analyzing Laboratory Testing Data	69
7.3.1	Test data for Wire Number 609	69
7.3.2	Test Data for Wire Number 613	71
7.3.3	Summary of all Test Data	72
7.3.4	Cracked Wires as a Separate Group	75
7.4	Cable Strength Evaluation – Simplified Model	75
7.4.1	Estimate of Number of Broken Wires in the Development Length (Panel)	76
7.4.2	Cracked Wires in the Evaluated Panel	77
7.4.3	Estimate of Cable Strength	78
8.0	Appendix B: Previous Inspection References	82
9.0	Appendix C: Flowcharts	83
9.1	Inspection Flowchart	83
9.2	Strength Evaluation Flowchart	84
10.0	Appendix D: Inspection and Evaluation Forms	85
10.1	Inspection Forms	85
10.2	Strength Evaluation Forms	93
11.0	Appendix E: BTC Method for Evaluation of Remaining Strength and Service Life of Bridge Cables	98
11.1	Introduction	98

11.2 Main Cable Inspection and Sampling	99
11.2.1 Panel Selection Criterion	99
11.2.2 Random Sampling and Sample Size Determination	99
11.2.2.1 Random Sampling and Practical Considerations	100
11.2.2.2 Sampling Size Determination	100
11.2.2.3 Wedge Pattern	101
11.2.2.4 Sampling Frame of Random Sample	101
11.2.3 Inspection Procedures	102
11.3 Wire Testing Program	104
11.3.1 Enhanced Tensile Strength Test in Standard Wire Specimens	104
11.3.2 Tensile Strength Test on Long Wire Specimens	104
11.3.3 Fracture Toughness Test	104
11.3.4 Fractographic and Scanning Electron Microscope (SEM) Evaluation	105
11.4 Cable Strength Evaluation	105
11.4.1 BTC Method Inputs	105
11.4.2 Choice of Probability Distributions	106
11.4.3 Elongation Threshold Criterion, $M_{threshold}$	106
11.4.4 Determination of Wire Condition	107
11.4.5 Wire Recovery Length	107
11.4.6 Broken Wires	108
11.4.6.1 Exterior Broken Wires	108
11.4.6.2 Interior Broken Wires	109
11.4.7 Cracked Wires	109
11.4.8 Strength Evaluation using the BTC Method	109
11.5 BTC Method Forecast of Cable Life	109
11.5.1 Forecast of Degradation in Intact Wire Strength	110
11.5.2 Forecast of Degradation in Cracked Wire Strength	110
11.6 Sensitivity Analysis	111
11.6.1 Key Inputs	111
11.6.2 Sensitivity Indices	112
11.7 Appendix E References	112

List of Figures

Figure 1-1 Drawing showing an elevation view of a typical suspension bridge (taken from *NCHRP Report 534*) 4

Figure 1-2 Drawings showing an elevation and cross-sectional view of a typical tower saddle (taken from *NCHRP Report 534*) 6

Figure 1-3 Drawing showing an elevation view of a cable anchorage (taken from *NCHRP Report 534*) 7

Figure 1-4 Drawing showing an elevation view of cable bent saddle (taken from *NCHRP Report 534*) 7

Figure 1-5 Photo showing a clamping collar and splay casting 8

Figure 1-6 Photo showing a cable band with no suspenders 9

Figure 1-7 Photo showing a cable band with suspenders 9

Figure 1-8 Inspection photo showing small perforations in the cable wrap which is a potential water ingress location 11

Figure 1-9 Photographs showing the four stages of cable wire corrosion [taken from [5]] 12

Figure 1-10 Photographs showing local pitting corrosion and pitting with localized exfoliation of iron oxide [taken from [5]] 13

Figure 2-1 Drawing of a three-span suspension bridge highlighting various components (taken from BIRM [3]) 14

Figure 2-2 Typical cable biennial inspection form (taken from *NCHRP Report 534*) 17

Figure 2-3 Drawing showing a protective sleeve adjacent to tower saddle (taken from *NCHRP Report 534*) 18

Figure 2-4 Drawing showing an elevation view and cross-section of a typical cable band (taken from *NCHRP Report 534*) 18

Figure 2-5 Drawing showing the typical components of an anchor block (taken from BIRM [3]) 19

Figure 2-6 Typical cable inside anchorage biennial inspection form (taken from *NCHRP Report 534*) 20

Figure 2-7 Form for recording defects in the suspender cable system (taken from BIRM [3]) ... 21

Figure 2-8 Photo showing damaged caulking and paint at a cable band (taken from *NCHRP Report 534*) 23

Figure 2-9 Photo showing a ridge which indicates crossing wires (taken from *NCHRP Report 534*) 23

Figure 2-10 Photo showing a hollow area which indicates crossing wires (taken from *NCHRP Report 534*) 24

Figure 2-11 Form for recording locations of internal cable inspections (taken from *NCHRP Report 534*) 25

Figure 2-12 Form for recording observed wire damage inside the wedged opening (taken from *NCHRP Report 534*) 27

Figure 2-13 Form for recording locations of broken wires and samples for testing (taken from *NCHRP Report 534*) 28

Figure 2-14 Photo showing a typical cable wedged open for inspection (taken from *NCHRP Report 534*) 29

Figure 2-15 Photograph of a cracked wire (taken from *NCHRP Report 534*).............................. 34

Figure 3-1 Typical form for recording observed wire deterioration... 39

Figure 3-2 Typical wire deterioration/corrosion map 40

Figure 3-3 Graph for computing the inverse of the standard normal cumulative distribution function (taken from *NCHRP Report 534*) 44

Figure 4-1 Graph for computing the strength reduction factor, *K* (taken from Figure 5.3.3.1.2-1 of *NCHRP Report 534*) 49

Figure 7-1 Drawing showing wires in half-sectors (taken from *NCHRP Report 534*)................. 57

Figure 7-2 Drawing showing a cable divided into eight sectors (taken from *NCHRP Report 534*) 58

Figure 7-3 Drawing showing field inspection data for the sector five wedged opening 61

Figure 7-4 Drawing showing the cable wire corrosion map... 63

Figure 7-5 Drawing showing the map of broken wires and wires removed for testing (taken from *NCHRP Report 534*) 65

Figure 7-6 Graph for computing the inverse of the standard normal cumulative distribution 71

Figure 7-7 Graph for computing the strength reduction factor, K (taken from Figure 5.3.3.1.2-1 of *NCHRP Report 534*) 80

Figure 10-1 Typical cable biennial inspection form (taken from Figure 2.2.3.1-1 of *NCHRP Report 534*) 85

Figure 10-2 Typical summary form showing biennial inspection rating system (taken from Figure 2.2.3.1-2 of *NCHRP Report 534*) 86

Figure 10-3 Typical form for biennial inspection showing detailed ratings (taken from Figure 2.2.3.1-3 of *NCHRP Report 534*) 87

Figure 10-4 Typical form for biennial inspection of cable inside anchorage (taken from Figure 2.2.3.2-1 of *NCHRP Report 534*) 88

Figure 10-5 Form for recording locations of internal cable inspections (taken from Figure 2.2.1.2.4-1 of *NCHRP Report 534*) 89

Figure 10-6 Form for recording observed wire damage inside wedged opening (taken from Figure 2.3.1.2.4-2 of *NCHRP Report 534*) 90

Figure 10-7 Form for recording locations of broke wires and samples for testing (taken from Figure 2.3.1.2.4-3 of *NCHRP Report 534*) 91

Figure 10-8 Form for recording cable circumference (taken from Figure 2.4.1-1 of *NCHRP Report 534*) 92

Figure 11-1 Eight-wedge pattern and pool of wire samples in the cable cross section 101

Figure 11-2 Photo showing the sample tag in wedge #7, wedge #8 side, ring #4, panel point 3-4 102

Figure 11-3 Photo showing wedges driven to the center of the cable, interior broken wire is shown 103

Figure 11-4 Photographs showing the four stages of corrosion 104

List of Tables

Table 2-1 Interval between internal inspections *(taken from NCHRP Report 534)* 22

Table 2-2 Sample lengths and number of specimens from each sample (taken from *NCHRP Report 534*) 32

Table 7-1 Number of wires in each ring 60

Table 7-2 Corrosion stages assigned to individual wires 62

Table 7-3 Number of wires in each corrosion stage 68

Table 7-4 Results of tension tests for wire number 609 70

Table 7-5 Results of tension tests for wire number 613 72

Table 7-6 Summary of all tension test results 73

Table 7-7 Cracked wires in the evaluated panel 78

Table 7-8 Calculations for mean tensile strength of the cable 79

Table 10-1 Typical table that can be used for assignment of corrosion stages to individual rings of wires 93

Table 10-2 Typical table that can be used for assignment of corrosion stages to half sectors of cable 94

Table 10-3 Typical table that can be used to record tension test results for a single wire 95

Table 10-4 Typical table that can be used to summarize the tension test results 96

Table 10-5 Typical table that can be used to summarize the number of broken and cracked wires 97

Table 10-6 Typical table that can be used for calculating the mean tensile strength 97

FOREWORD

The motivation for the development of the "Primer for the Inspection and Strength Evaluation of Suspension Bridge Cables" is to provide a practical supplement to the NCHRP Report 534, "Guidelines for Inspection and Strength Evaluation of Suspension Bridge Parallel Wire Cables," and the FHWA "Recording and Coding Guide for the Structure Inventory and Appraisal of the Nation's Bridges." This Primer serves as an initial resource for planning and performing inspection, metallurgical testing, and strength evaluation of suspension bridge cables. This Primer also provides an example of a simplified strength evaluation, flowcharts illustrating the inspection and strength evaluation procedures, and inspection and strength evaluation forms that can be used, or replicated, by inspectors and engineers.

Suspension bridges are significant investments in our nation's infrastructure, in addition to serving as public lifelines. In many cases, suspension bridges are essential transportation links for regional, national, and international commerce. As these key infrastructure investments advance in age, there is a need to efficiently inspect and evaluate the strength of these bridges to ensure that they have adequate load-carrying capacity. Furthermore, bridge owners will desire to maintain and extend the service life of these bridge types based upon standardized inspections and strength evaluations.

The Primer is expected to be of immediate interest to suspension bridge field inspectors, technicians, laboratory personnel, bridge engineers, and bridge owners.

The constructive review comments on the final draft provided by many engineering professionals are very much appreciated. The readers of this Primer are encouraged to submit comments for enhancements of future editions of the Primer to Myint Lwin at the following address: Federal Highway Administration, 1200 New Jersey Avenue, S.E., Washington D.C. 20590.

M. Myint Lwin, Director
Office of Bridge Technology

ACKNOWLEDGEMENTS

The authors would like to acknowledge the encouragement and guidance provided by Myint Lwin, the Director of the Office of Bridge Technology, and Raj Ailaney, the Contract Manager of the Office of Bridge Technology during the development of the *Primer*. The authors also would like to thank the engineering professionals that provided very useful review comments on the final draft of the *Primer*.

1.0 INTRODUCTION

The intent of this Primer is to supplement NCHRP Report 534, *Guidelines for Inspection and Strength Evaluation of Suspension Bridge Parallel Wire Cables* (Mayrbaurl and Camo 2004) [1][1], and FHWA Report No. FHWA-PD-96-001, titled *Recording and Coding Guide for the Structure Inventory and Appraisal of the Nation's Bridges* [2]. This primer will serve as an initial resource for those involved in the inspection, metallurgical testing, and strength evaluation of suspension bridge cables in addition to providing necessary documentation for recording performed inspections, testing, and strength evaluations. Furthermore, this document is intended to provide field inspectors, technicians, and/or engineers with the necessary forms and information they need to perform an inspection.

1.1 DOCUMENTS

The *Guidelines for Inspection and Strength Evaluation of Suspension Bridge Parallel Wire Cables*, herein referred to as *NCHRP Report 534*, is an in-depth resource that can be used for assessing cable integrity through inspection, metallurgical testing, and strength evaluation. The Report also provides a method of standardization for the process of evaluating a cable that has been in service for an extended period of time. This Primer will highlight the critical aspects of cable inspection, laboratory testing, strength evaluation, and documentation of the entire process. Much of the information provided in this Primer is taken directly from the Report. However, for a more detailed treatment of the subjects contained in this Primer as well as additional topics, the reader is encouraged to review *NCHRP Report 534*.

The Recording and Coding Guide for the Structure Inventory and Appraisal of the Nation's Bridges provides detailed guidance in evaluating and coding specific bridge data that comprise the National Bridge Inventory database. This guide was developed for use by the states, federal and other agencies so as to have a complete and thorough inventory of the nation's bridges.

1.2 PRIMER ORGANIZATION

Section 2 of this Primer provides guidelines for inspecting suspension bridge parallel wire cables as detailed in *NCHRP Report 534* and supplemented from Section 12, Special Bridges, Topic 12.1 Cable Supported Bridges in Publication No. FHWA NHI 03-001, *Bridge Inspector's Reference Manual* [3]. Laboratory testing methods typically employed for suspension bridge parallel wire cables are as discussed in Section 2. An overview of the tabulation and presentation of field and laboratory observations is provided in Section 3. In Section 4, the methods that can be used to estimate the strength of the suspension cables are presented and discussed. These strength evaluation methods include the Simplified Model, the Brittle-Wire Model, and the Limited Ductility Model. Section 5 explains the documentation, reporting, and recommendations that should be created after the inspection and evaluation of a suspension bridge cable, which will allow owners to make informed decisions about maintenance schedules and budgets.

---

[1] Numbers in brackets refer to References provided in section 6.0 of this document.

References are provided in Section 6, while Section 7 (Appendix A) contains an in-depth strength evaluation example using the Simplified Model. A list of references for inspection and evaluation projects that have employed the criteria of *NCHRP Report 534* is provided in Section 8 (Appendix B). Section 9 (Appendix C) contains flowcharts demonstrating the processes for cable inspection and strength evaluation. Section 10 (Appendix D) includes blank forms that can be used by cable inspections, as well as tables that can be followed to perform strength evaluation calculations associated with the Simplified Model.

Section 11 (Appendix E) presents the BTC Method, an alternative methodology to that provided in the *NCHRP Report 534*, for the assessment of remaining strength and residual life of bridge cables. The method applies to both; parallel and helical; either zinc-coated or bright wire suspension and cable-stayed bridge cables. The BTC method includes random sampling without regard to wire appearance, mechanical testing of wire samples, determining the probability of broken and cracked wires, evaluating ultimate strength of cracked wires employing fracture mechanics principles and utilizing the above data to assess remaining strength of the bridge cable in each panel. The probabilistic-based method forecasts remaining service life of the cable by determining the rate of growth in proportions of broken and cracked wires over a time frame, measuring the rate of change in effective fracture toughness over same time frame, and applying the rates of change to a strength degradation prediction model. The BTC method provides a sensitivity analysis to identify the key inputs which influence the estimated cable strength and assist in decision making regarding future maintenance. Persons interested in using the BTC method should contact the author of this Appendix.

1.3 SUSPENSION BRIDGE CABLES

Suspension bridges are large, unique structures with two or more cables that carry the weight of the deck and the imposed live load to the towers that support the cables. The suspension cables are in tension and require massive anchorage at both ends, and are typically load-path nonredundant. Figure 1-1 shows an elevation view of a typical suspension bridge with main bridge components labeled.

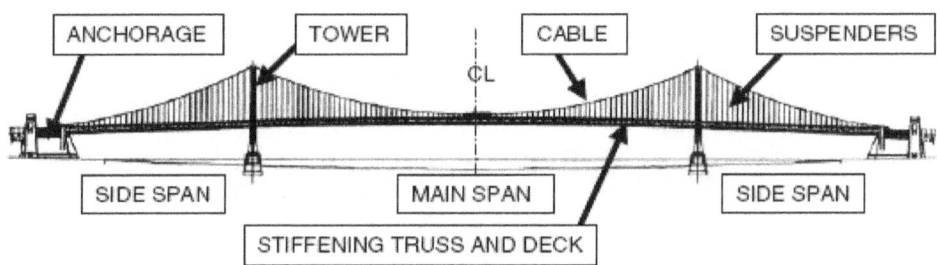

Figure 1-1 Drawing showing an elevation view of a typical suspension bridge (taken from *NCHRP Report 534*)

The cables are constructed of many individual wires, typically laid parallel to one another and clamped at points where suspenders connect with them to support the bridge deck. For most North American bridges, these individual wires have a 0.192 inch diameter, and a 0.002 inch zinc coating around the wire, resulting in a total diameter if 0.196 inch. Bridge cable wires are typically coated with zinc to provide cathodic protection to protect the wire steel against corrosion. The quality of the zinc coating is important to ensure wire safety; discontinuities produced during manufacturing or installation can facilitate corrosion of the steel wire.

Suspension bridge cable wires are made of ultra high strength steels because of the heavy loads they are required to support. ASTM A586 mandates that the bridge cable wires are to have a minimum tensile strength of 220 ksi (kips per square inch). In some cases, modern wires exceed this specification with tensile strengths as high as 260 ksi. The requirement of 220 ksi strength is based on the gross metallic area, including the zinc coating.

It should be noted that for the strength evaluation of older bridges, employing the methodology of this Primer and *NCHRP Report 534*, the above discussion concerning material properties may not be appropriate. In some cases, bridge cable wires may have a minimum tensile strength well below 220 ksi. The bridge engineer performing the inspection and evaluation must be aware of the original material specifications that may apply to the structure being investigated, or should otherwise have the particular component tested to determine necessary material properties.

1.3.1 Bridge Cable Components

The performance of the bridge cable wires and their inspectability are affected by additional variables other than the cables themselves. Details such as tower saddles, cable bent saddles, splay castings, cable anchorages, and the connection of the suspenders to the cables can affect the performance and inspectability of the cable wires.

The vertical forces at the top of the towers are transferred from the suspension bridge cable into the tower by the tower saddles. In most cases, the entire weight of the bridge is supported at the top of the tower. A typical tower saddle is shown in Figure 1-2.

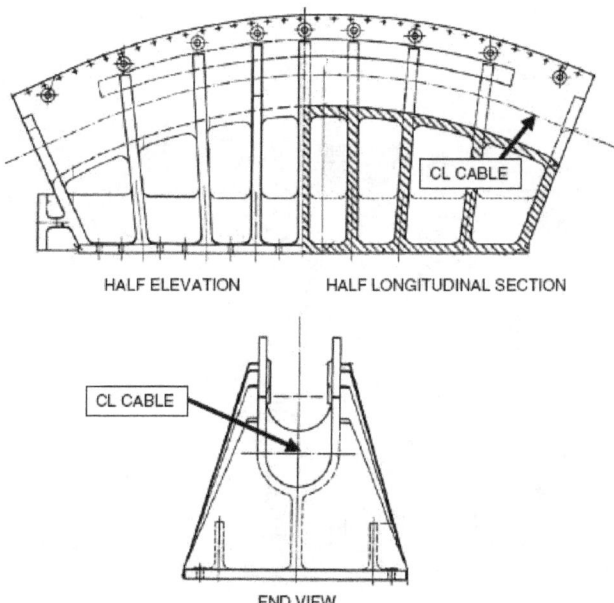

Figure 1-2 Drawings showing an elevation and cross-sectional view of a typical tower saddle (taken from *NCHRP Report 534*)

Anchorage and cable bent saddles bend the suspension bridge cables so that they come into alignment with the anchoring mechanisms. Anchorage saddles sometimes have a variable vertical radius and horizontal flare, allowing the cable strands to be splayed directly to their anchoring mechanisms. A typical anchorage is shown in Figure 1-3.

Cable bent saddles typically bisect the interior angle of the cable, so that the cable changes direction within a single vertical curve from the saddle toward the anchorage without flaring to a splay casting. Cable bent saddles are typically supported on independent struts that may be hinged at the base or, if flexible enough, fixed at the base. A typical cable bent saddle is shown in Figure 1-4.

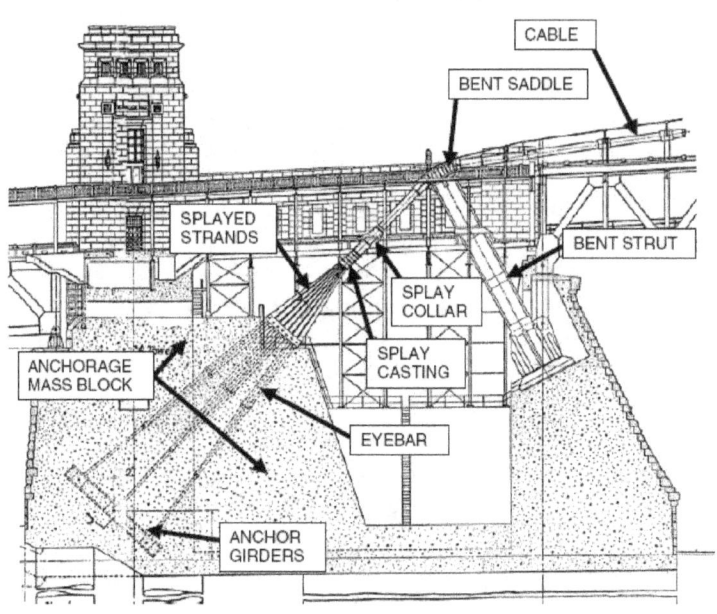

Figure 1-3 Drawing showing an elevation view of a cable anchorage (taken from *NCHRP Report 534*)

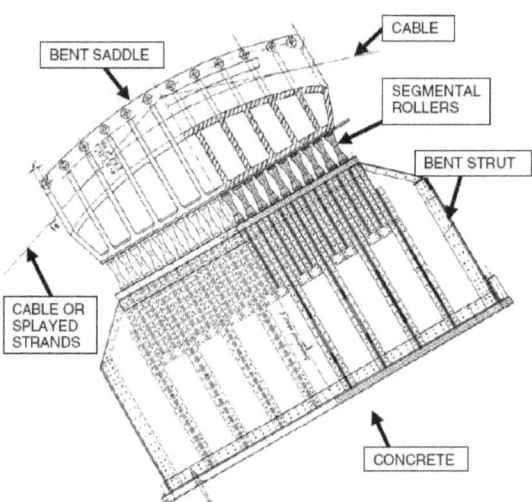

Figure 1-4 Drawing showing an elevation view of cable bent saddle (taken from *NCHRP Report 534*)

Splay castings, as shown in Figure 1-5, are used to control the direction of the strands that flare out of their respective anchoring devices. Castings resists the outward forces exerted by the strands, and are anchored against upward slippage by a cable collar clamped above the splay

7

casting. Corrosion of the wires inside the splay casting may be caused by water passing through the cable. However, inspecting wires inside the splay casting is a complex operation that typically requires a temporary relocation of the splay casting.

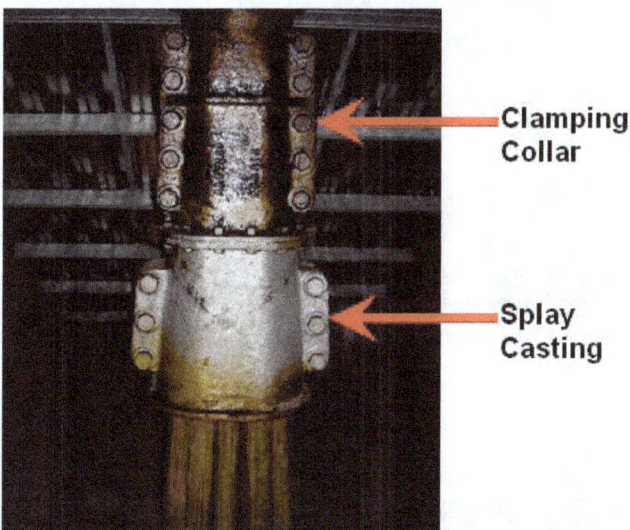

Figure 1-5 Photo showing a clamping collar and splay casting

Cable anchoring devices can consist of strand shoes, parallel wire strand terminations, or eyebars. Traditionally, cable wires loop around strand shoes, which are anchored to two eyebars by a pin. Alternatively, a single eyebar can be used with two strand shoes, one at each end of the pin, splitting the strand into four quarters. The strand shoe can also be restrained with high-strength strengthening rods. When a humid atmosphere exists in the anchorage or water enters the anchorage area, corrosion is often found in the lower half of the strands, particularly at the interfaces of the wires and the strand shoe where water tends to collect. Parallel wire strand terminations use zinc or polyester thermoset resin sockets rather than strand shoes, and the sockets are connected to anchoring assemblies embedded in the anchorage concrete. Eyebars are anchored to a grillage buried deep in the concrete mass of the anchorage. The focus of the eyebars may be slightly beyond the splay casting or cable bent saddle. To prevent eyebars from bending, spacers are placed between the eyebars of each separate strand so that the eyebars bear against each other. In humid anchorages, eyebar corrosion is typically found at the interface with the concrete mass, and is often hidden behind pack rust.

Cable bands consist of two cylinder halves bolted together over the circumference of the cable. The number of bolts per cable band is dependent on the slope of the cable at the suspender attachment point. The friction from squeezing against the cable prevents the cable band from sliding down the cable. More bolts are needed as the cable becomes steeper to prevent the cable band from sliding. Figure 1-6 shows a cable band without suspenders attached, and Figure 1-7 shows a cable band with suspenders attached.

Figure 1-6 Photo showing a cable band with no suspenders

Figure 1-7 Photo showing a cable band with suspenders

The wires in a suspension cable are often protected by wire wrapping, which typically consists of soft galvanized No. 9 wires with Class A zinc coating. Some newer bridges have used an S-shaped wire that interlocks with the other wires. Wrapping is installed by power-driven machines with multiple reels that are capable of placing from one to three wires at a time. The wrapping wires are in a single layer, in side-by-side helices. Paint systems are used to cover and seal the wrapping wires. Other protection systems employ elastomeric membranes or fiberglass reinforced lucite composites and methacrylates.

1.3.2 Bridge Cable Protection

The individual cable wires can be protected with the use of zinc coatings, grease and oil, and/or paste mixtures, while the complete suspension cable is typically protected with galvanized wire wrapping and paint. With few exceptions, the cable wires are protected with a zinc coating, which can last indefinitely or could become defective within 20 years depending on the effectiveness of the exterior protective system. The zinc coating provides cathodic protection, and exploits the phenomenon of galvanic action to protect the steel against corrosion. In this application, the zinc coating provides cathodic protection by depleting in the presence of water thus protecting the steel.

Some early bridge cable wires were greased during spinning or as the cable was being compacted. In some cases the greased wires have appeared almost new after many years of service and in other cases, despite the presence of grease, wires were known to crack and fail in localized zinc depleted regions.

Furthermore, various paste mixtures were used as a layer of protection under wrapping wires to prevent water penetration. In the past, red lead paste was traditionally used under the wrapping wire; however, given the hazardous nature of red lead, zinc-based pastes have been used in Europe and the United States. Membrane protection and dehumidification are now widely used in Asia and have been used in Europe.

1.3.3 Causes of Cable Deterioration

The deterioration of suspension bridge cable wires is principally caused by corrosion. Corrosion is caused by the presence of water and its solutes. There are several factors that affect a cable's susceptibility to corrosion, including environmental aspects, amount of water penetration, installation practices, and the vulnerability of the wires to corrosion attack.

The term macro-environment can be used to describe the environmental conditions that affect the structure as a whole. A suspension bridge's macro-environment often contains moisture, pollutants, dissolved gases, and salt spray from deicing salts or coastal environment, all of which may contribute to corrosion of the cable wires.

The term micro-environment can be used to describe the conditions inside the cable that affect the individual wires. Water can enter a cable as a liquid from either precipitation or as a vapor during periods of high temperature and humidity. The water vapor will turn to liquid form as the temperature falls, forming condensation on the wire surfaces. Some micro-environments, which can act alone or together, observed in bridge cables are:
- Acid rain chemistry, leading to hydrogen evolution from the reaction with the zinc wire coating
- Carbonate or bicarbonate chemistry, either alkaline or highly acidic
- Nitrate chemistry, either alkaline or acidic
- Alkaline chemistry
- Seawater or salt spray, moderately acidic
- Cathodic action in which a metal more noble than steel is placed in contact with the wires

These micro-environments can cause cable wires to corrode, crack, and/or break.

There are several methods in which water can penetrate into the cable. A breakdown in the exterior protection system, such as poorly wrapped cables or cracks in the paint, can allow water to enter the cable. Furthermore, perforations can develop in the cable wrapping, as shown in Figure 1-8. Joints on the underside of the cable are often provided to allow for weeping of internal water. However, these joints may become points of entry for water streaming along the underside of the cable or from wind-driven rain. Damaged or poorly maintained housings for saddles and anchorages can allow water to enter the cable or cause damage to the wires near the saddles and anchorages. Lastly, paint cracks and other entry points for water ingress are also entry points for water vapor, which can lead to condensation in the cables.

Figure 1-8 Inspection photo showing small perforations in the cable wrap which is a potential water ingress location

The cable installation practices can lead to deterioration of the wires. Poor cable compaction and crossing of the wires can cause unusually large voids in the cable that allow water to penetrate deep into the cable. The crossing of wires can also expose steel at the point of contact, and therefore accelerate cathodic action.

The individual wires are more susceptible to corrosion than milled rolled steel due to the processes used to fabricate the wires, which includes a high carbon content and cold working of the steel. The zinc coating used to protect the wires from corrosion is beneficial as long as there are no breaches in the coating. If the zinc coating is damaged or missing, corrosion is more likely to occur.

The "cast" of the wire, or its natural curvature (on the order of 4ft in diameter), which was necessary to initially spin the cables, has inherent high residual stress and very high straightening stresses predisposing the wires to be attacked in the inside radius of the wire. This is the foremost culprit in wire damage. Modern specifications call for as large of a "cast" as possible, where a diameter on the order of 30ft is possible. Wires manufactured with small cast radii have a high residual stress, estimated to be 30 to 36 ksi by X-ray diffraction.

1.3.4 Cable Wire Corrosion

The corrosion mechanism for zinc-coated cable wires within the span is different than mechanisms for zinc-coated cable wires in the anchorages and for uncoated wires. The discussion that follows concerns cable wires within the span. Through visual inspection, wire corrosion is classified into four different stages, developed by Hopwood and Havens [4]. The classification system has over time provided accurate descriptions of the various stages of

corrosion in the cable wires, and produced a usable grouping for strength evaluation of the cable. As shown in Figure 1-9, the four stages of wire corrosion that are typically used are:
- *Stage 1*: white spots on the surface of the wire, indicating early stages of zinc oxidation
- *Stage 2*: white zinc oxidation over the entire wire surface
- *Stage 3*: white zinc oxidation in some areas of the wire, with brown rust spots not covering more then 30% of a 3 in. to 6 in. length of wire
- *Stage 4*: brown spots prevalent over the wire surface, covering more than 30% of a 3 in. to 6 in. length of wire

A 5th stage is often used as well, to represent wires that have stage 4 corrosion (above), but with cracks in the wires.

A Stage 2 wire may have white surface dust, indicating zinc oxidation, but it does not necessarily imply depletion of the zinc coating. Depletion of the zinc coating is typically indicated by a dull gray color, or a dark gray to black color if sulfur is present.

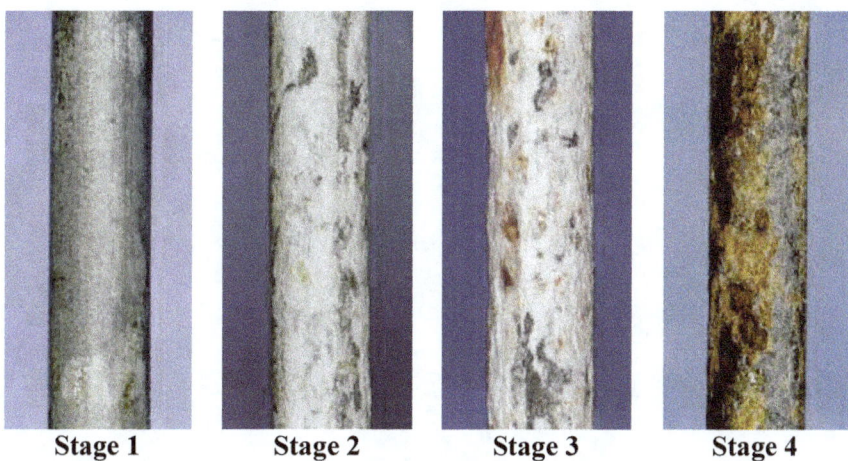

Stage 1   Stage 2   Stage 3   Stage 4

Figure 1-9 Photographs showing the four stages of cable wire corrosion [taken from [5]]

In addition to surface corrosion, pits of various types can be found in the wires. As shown in Figure 1-10, some Stage 4 wires can show extensive exfoliation and/or local pitting, or pitting characterized by highly localized exfoliation of iron oxide. Furthermore, laboratory tests have shown that 5% to 20% of Stage 3 corroded wires, and 60% of Stage 4 corroded wires, may have cracks.

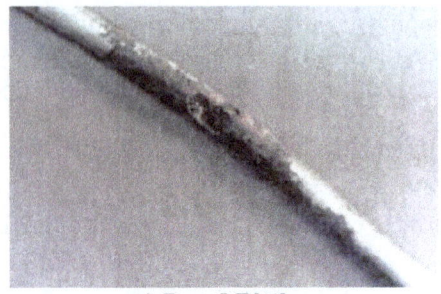 

a) Local Pitting   b) Pitting with Localized Exfoliation of Iron Oxide

Figure 1-10 Photographs showing local pitting corrosion and pitting with localized exfoliation of iron oxide [taken from [5]]

2.0 INSPECTION GUIDELINES AND LABORATORY TESTING METHODS

This section will discuss guidelines for inspecting suspension bridge parallel wire cables as detailed in *NCHRP Report 534* and supplemented from Section 12, Special Bridges, Topic 12.1 Cable Supported Bridges in Publication No. FHWA NHI 03-001, *Bridge Inspector's Reference Manual* (BIRM) [3]. Laboratory testing methods and results, used to estimate wire strength and ultimately evaluate cable strength, will also be highlighted herein.

2.1 CABLE INSPECTION

A typical suspension bridge is comprised of a deck system that is connected by vertical suspender cables to the main suspension cables, which are generally supported on saddles atop towers and are anchored at both ends. Figure 2-1 depicts a three-span suspension bridge schematic, identifying these major components. Suspension bridges with only two main suspension cables offer only two load paths (non-redundant); therefore, the two tension cables are identified as fracture critical members for the purpose of inspection and evaluation.

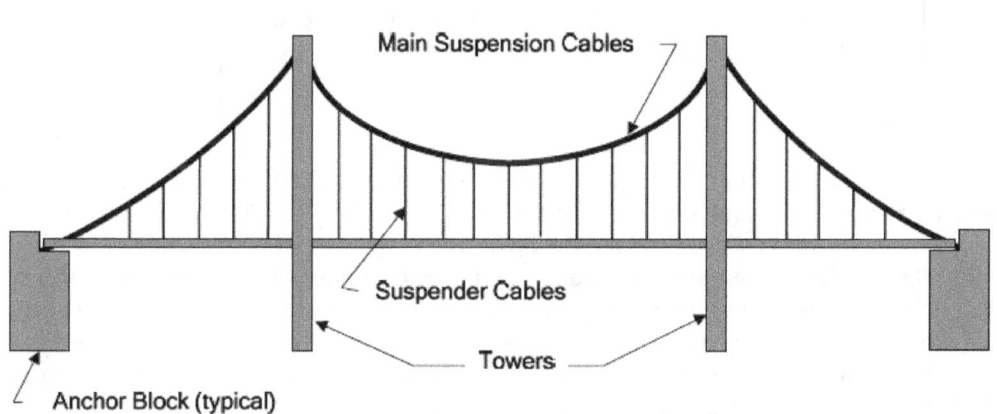

Figure 2-1 Drawing of a three-span suspension bridge highlighting various components (taken from BIRM [3])

The goal of the cable inspection is to obtain information about the condition and strength of the cable wires, which can then be used to evaluate suspension bridge cable strength. Although several levels of inspection are performed over the lifespan of the structure, only internal cable inspections provide data for strength evaluation.

Cable inspections should be led by a chief inspector, a professional engineer with experience in bridge cable inspections. Cable inspection over trafficked roadways and waterways involves risk to people and the environment. Protection of construction workers, inspectors, motorists, pedestrians, and marine traffic is an important consideration. Personnel associated with the cable inspection should understand health hazards and be trained in the use of equipment (full-body safety harness, dual shock-absorbing lanyards, etc.) and monitoring procedures (blood-lead baseline and subsequent checks for lead absorption) associated with health maintenance.

Reference the OSHA Compliance Manual Training requirements and Subsection 1.2 Health and Safety Requirements of *NCHRP Report 534*.

2.1.1 Levels of Inspection and Inspection Intervals
There are three levels of cable inspection:
- Periodic routine visual inspections by maintenance personnel of the cable exterior
- Biennial hands-on inspections of non-redundant members
- Scheduled thorough internal inspections

2.1.1.1 Inspections by Maintenance Personnel

Periodic inspection tours of the cable by maintenance personnel are recommended, beginning by inspecting the underside of the cable with binoculars, and then walking the cable along its full length. These inspections should occur at the end of winter (March or April) to observe damage due to frost or deicing salts in the splash zone, and at the end of summer (September or October) to observe the effects of extreme heat on paint and caulking. Additional inspections should occur after severe snow, ice, rain or wind storms. During these inspections, the underside of the cable should be examined for evidence of water penetration (dripping from the wrapping wire or weep holes in the lower cable band grooves, and unusually damp areas).

2.1.1.2 Biennial Inspections

In accordance with the National Bridge Inspection Standards (NBIS), non-redundant fracture critical members receive hands-on inspections every 24-months. Suspension bridges are considered complex bridges according to the NBIS regulations. The NBIS requires identifying specialized inspection procedures, and additional inspector training and experience to inspect these complex bridges. Suspension bridges are to be inspected according to these procedures. Specifically developed bridge maintenance manuals, if available, should be used to verify that specified routine maintenance has been performed. Customized, preprinted inspection forms should be used to report findings in a systematic manner.

Cables in Suspended Spans: Inspect the main suspension cables for indications of corroded wires. Inspect the protective covering or coating, especially at low points of cables, areas adjacent to the cable bands and saddles over towers. The conditions of the following bridge components should be reported (see Figure 2-2 for sample form):
- Paint or Surface Protection, inspected for dried out, peeling, cracked and crazed paint
- Elastomeric Barrier (see Figure 2-3), inspected for puncture or tearing
- Caulking at Cable Bands, inspected for gaps or cracks
- Hand Ropes and Stanchions, inspected for broken wires, tightness and corrosion
- Wire Wrapping, inspected for anomalies including:
    - Unequal tension of wire plies, indicated by unevenness in wrapping surface
    - Bunching below or separating above the cable bands
    - Gaps in wrapping, corroded or broken wrapping wire
    - Surface ridges, indicated by crossing wires and hollow areas

- Saddles, inspected for missing or loose bolts, damaged sleeves, bellows or flashing, and corrosion or cracks in the casting. Check for proper connection to top of tower or supporting member, and possible slippage of the main cable.
- Cable Bands (see Figure 2-4), inspected for missing or loose bolts, rust stains or dripping water, indicative of internal corrosion, or broken suspender saddles. Check for the presence of cracks in the band itself as well as corrosion or deterioration of the band.
- Measure and report the cable diameter at several intervals along the cable. Later, the diameter of the cable in combination with the known number of wires, can be used to estimate the potential for water accumulation.

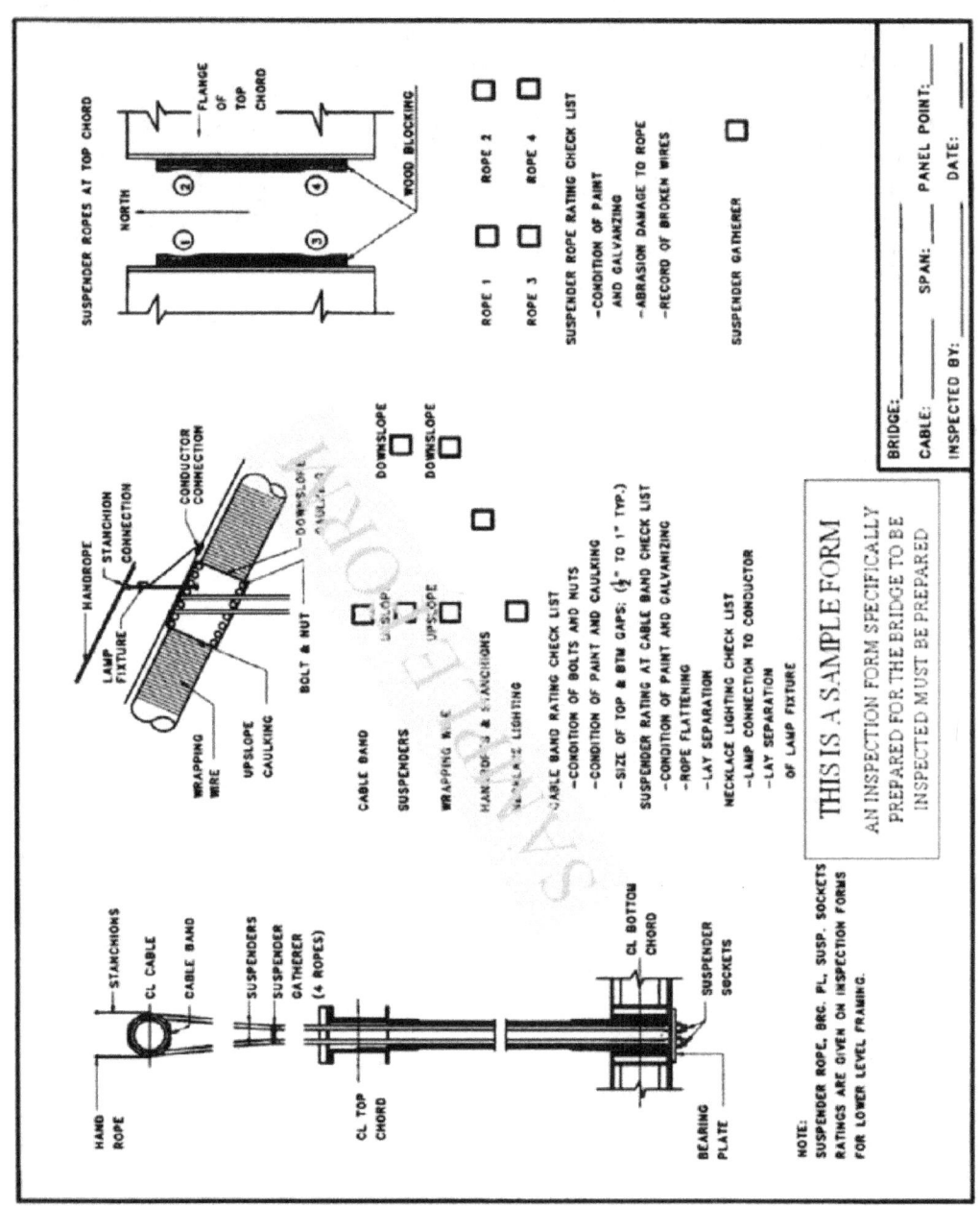

Figure 2-2 Typical cable biennial inspection form (taken from *NCHRP Report 534*)

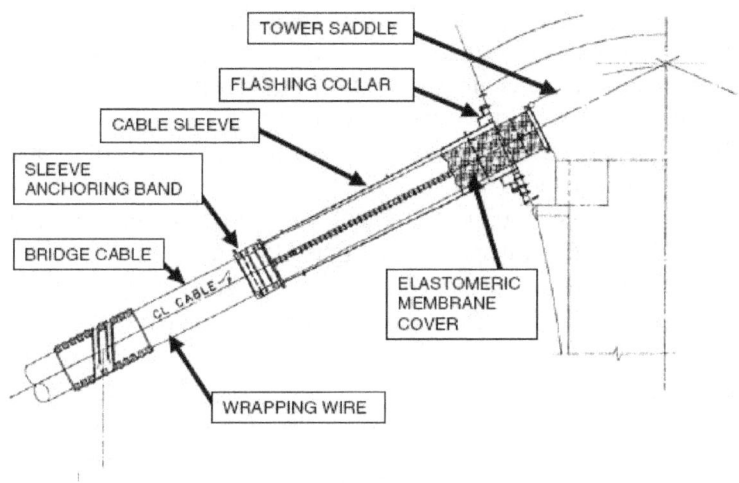

Figure 2-3 Drawing showing a protective sleeve adjacent to tower saddle (taken from *NCHRP Report 534*)

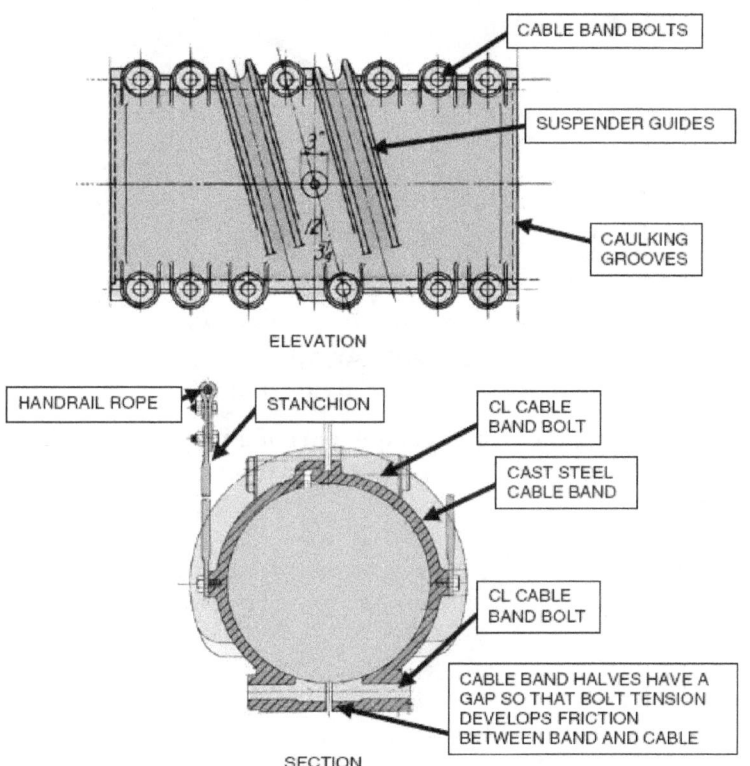

Figure 2-4 Drawing showing an elevation view and cross-section of a typical cable band (taken from *NCHRP Report 534*)

Cables Inside Anchorages: Inspect the anchorage system (see Figure 2-5 for schematic of components) at the ends of the main suspension cables. The splay saddle, bridge wires, strand shoes or sockets, anchor bars, and chain gallery need to be inspected. The conditions of the following bridge components should be reported (see Figure 2-6 for sample form):
- Strands Inside Anchorages, inspected for corrosion or broken wires, and swelling or bulges at the strand shoes
- Anchorage Walls and Roof (Chain Gallery), inspected for signs of water intrusion
- Eyebars and Strand Wires, inspected for signs of condensation
- Points of contact between Eyebars and Concrete Mass, inspected for corrosion
- Eyebars and Anchorage Strands, inspected for paint anomalies

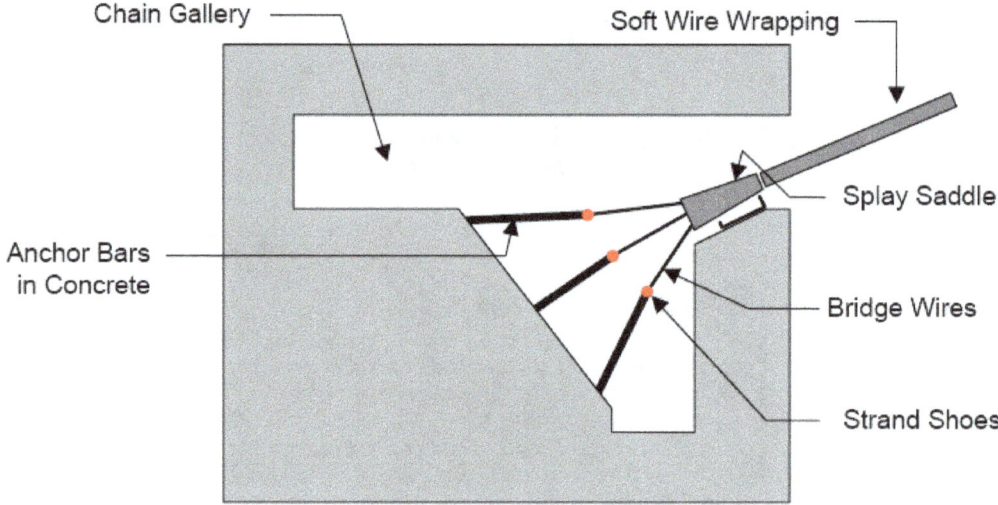

Figure 2-5 Drawing showing the typical components of an anchor block (taken from BIRM [3])

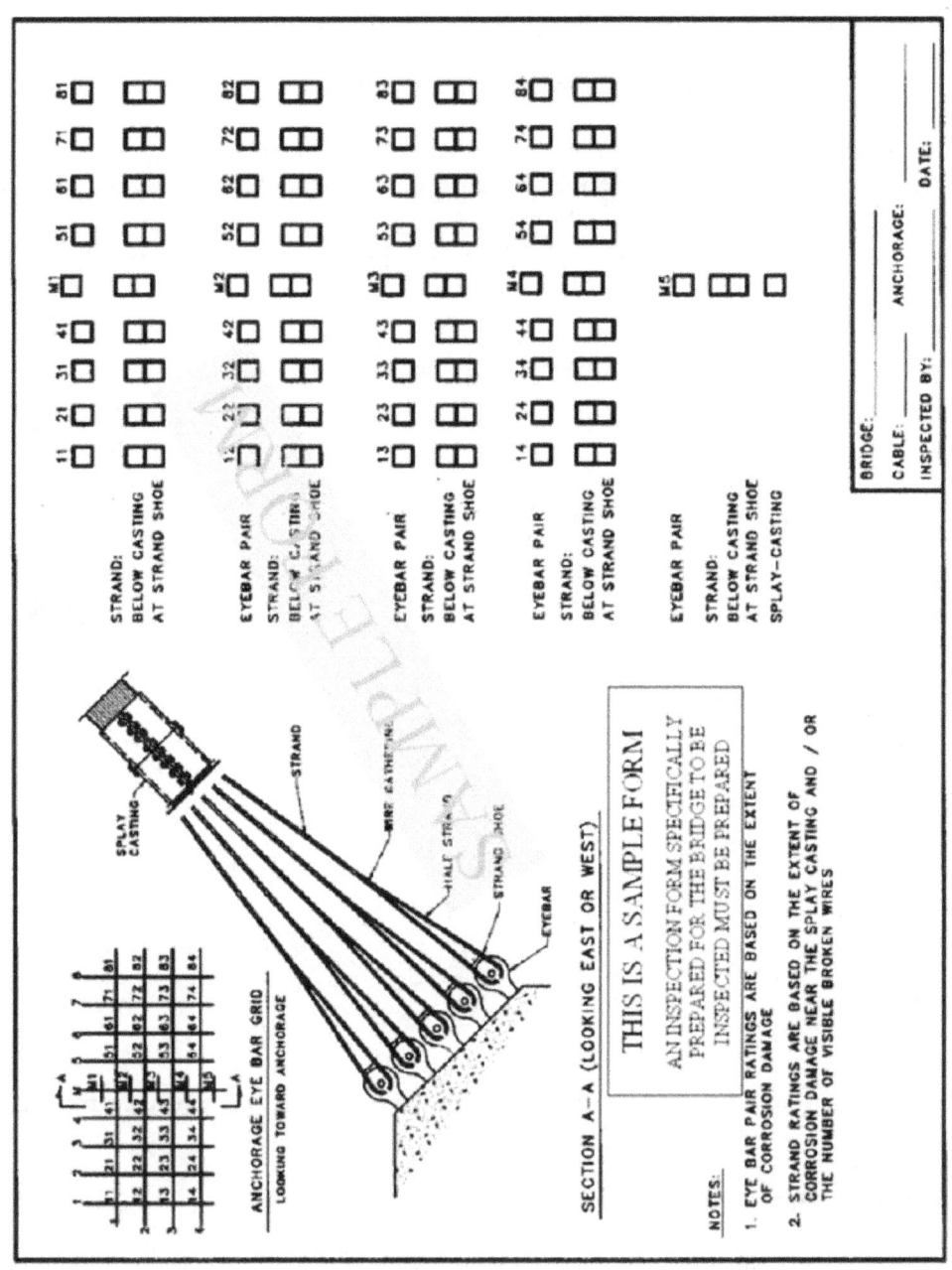

Figure 2-6 Typical cable inside anchorage biennial inspection form (taken from *NCHRP Report 534*)

Additional Components in Suspended Spans: Inspect the additional components attached to the main suspension cables. The suspender cables and connections, as well as sockets, need to be inspected. The conditions of the following bridge components should be reported (see Figure 2-7 for sample form):
- Suspender Cables and Connections, inspected for corrosion or deterioration, broken wires, and kinks or slack. Check for abrasion or wear at sockets, clamps and spreaders. Note excessive vibrations.
- Sockets, inspected for corrosion, cracks, deterioration and possible movement, or abrasion at connection to bridge superstructure

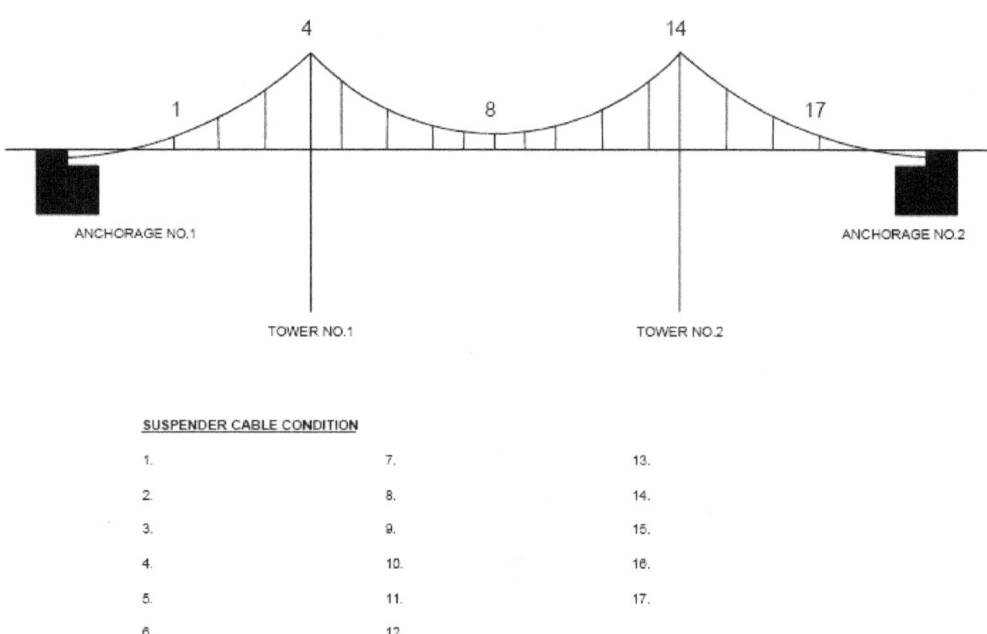

Figure 2-7 Form for recording defects in the suspender cable system (taken from BIRM [3])

2.1.1.3 Internal Inspections

Internal inspections are necessary at some point during the life of a cable. Suggested intervals between internal inspections are shown in Table 2-1. A baseline internal inspection of the cable should be performed when it has been in service for 30 years. Access to internal wires requires removing the cable's external protective system. At the discretion of the owner and the investigator, the suggested intervals could be adjusted based on the history of past internal inspections of the bridge cable (e.g. the presence of dissimilar metals such as copper or bronze in contact with, or in close proximity to, the wires, local deterioration from traffic collisions, or overheating the wires during maintenance operations). In addition, the interval between internal inspections should be shortened to 5 years when Stage 4 corrosion is found in more than 10% of the wires in the cable.

Table 2-1 Interval between internal inspections *(taken from NCHRP Report 534)*

Inspection Number	Maximum Corrosion Stage Found in Previous Inspection*	Age of Bridge at Last Inspection (Years)	Interval (Years)
First			30
Additional	1-(2)	any age	30
	2-(3)	40 or more	20
	2-(3)	30	10
	3-(4)	60 or more	20
	3-(4)	less than 60	10
	4	any age	10
	broken wires	any age	5

* Each corrosion stage may include up to 25% of the surface layer wires in the next higher stage, indicated by the number in parentheses. Stage 4 may include 5 broken surface layer wires.

Locations of Internal Inspections: Internal inspections should be located where external indications of deterioration are found. External signs of possible internal deterioration include: loose wrapping, dripping water from cable interior, rust stains, damaged caulking at cable bands (see Figure 2-8), surface ridges indicative of crossing wires underneath the wrapping (see Figure 2-9), or hollow sounding when "sounded" with a hammer (see Figure 2-10).

Figure 2-8 Photo showing damaged caulking and paint at a cable band (taken from *NCHRP Report 534*)

Figure 2-9 Photo showing a ridge which indicates crossing wires (taken from *NCHRP Report 534*)

Figure 2-10 Photo showing a hollow area which indicates crossing wires (taken from *NCHRP Report 534*)

If there are no signs of internal deterioration, the locations for internal inspections should be selected as follows (see Figure 2-11 for typical form for recording inspection locations):

- First Internal Inspection – A minimum of three locations along each cable should be selected as follows:
    - One in each cable at a low point of the Main Span
    - One in each cable at or near a low point of the Side Span
    - One in the first cable of the Main Span, above the low point at a distance from 30% to 70% of half the Main Span
    - One in the other cable of the Side Span, above the low point at a distance from 30% to 70% of the Side Span

The cables should be opened at each location (typically a panel length) and wedged at four locations around the perimeter. This should facilitate removal of, at the very least, a 10-foot long sample of wire from the outer two layers for testing. If the wire corrosion exceeds Stage 2, the opening should be extended to a full panel length, and wedged at eight locations around the perimeter. This will enable the driving of wedges to sufficient depth to determine the extent of Stage 3 or worse corrosion. It should be noted that when large numbers of broken or loose wires are found, the break often occurs within the cable band. All loose wires should be traced to a wire break, and therefore it may be necessary to temporarily remove a suspender and cable band to make this assessment.

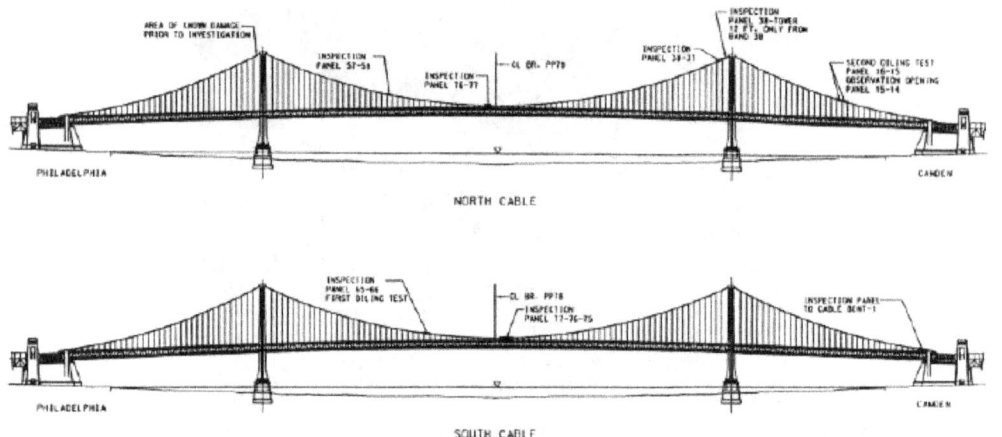

Figure 2-11 Form for recording locations of internal cable inspections (taken from *NCHRP Report 534*)

- Second Internal Inspection – The locations will be dependent upon the conditions found during the first internal inspection, as follows:
  - Stage 1 or Stage 2 corrosion revealed during the first internal inspection - A minimum of three locations along each cable should be selected following the logic of previous choices. The low point in the Main Span should be adjacent to the low point previously inspected. The Side Span location should be in the Side Span opposite the one previously inspected. One location in the Main Span and one location in a Side Span above the low points should also be inspected. The wedging protocol listed above should be followed herein (see Figure 2-12 and Figure 2-13 for typical forms for recording observed wire damage inside wedged openings, and locations of broken wires and samples for testing, respectively).
  - Stage 3 or Stage 4 corrosion to a depth of three wires or less revealed during the first internal inspection – Each cable should be internally inspected at six locations, including any one of the three previously inspected panels that exhibited Stage 2 corrosion or greater, and three additional locations recommended for the first internal inspection. Locations that exhibited only Stage 1 corrosion in the first internal inspection need not be reopened, but additional locations above the low points should be selected to bring the total locations to six. All six locations should be inspected for the full-length between cable bands, with wedges driven as deeply as possible to the center of the cable. Whenever Stage 4 corrosion is present to a depth greater than one wire, and the center of the cable cannot be reached with a full panel length unwrapped, one cable band per cable should be removed to adequately assess the condition of wires at the center of the cable.
  - Stage 4 corrosion to a depth of more than three wires – A minimum of 16% (preferably 20%) of the panels in each cable should be inspected internally. Four low points and two locations near the towers should be inspected; the balance of

locations should be selected at random in the remainder of the cable between the low points and the towers, one each from contiguous groups of panels that are approximately equal in number. The full-length of panels between cable bands should be inspected, with wedges driven as deeply as possible to the center of the cable. A minimum of two cable bands should be removed to facilitate inspection to the center of the cable and under the bands.

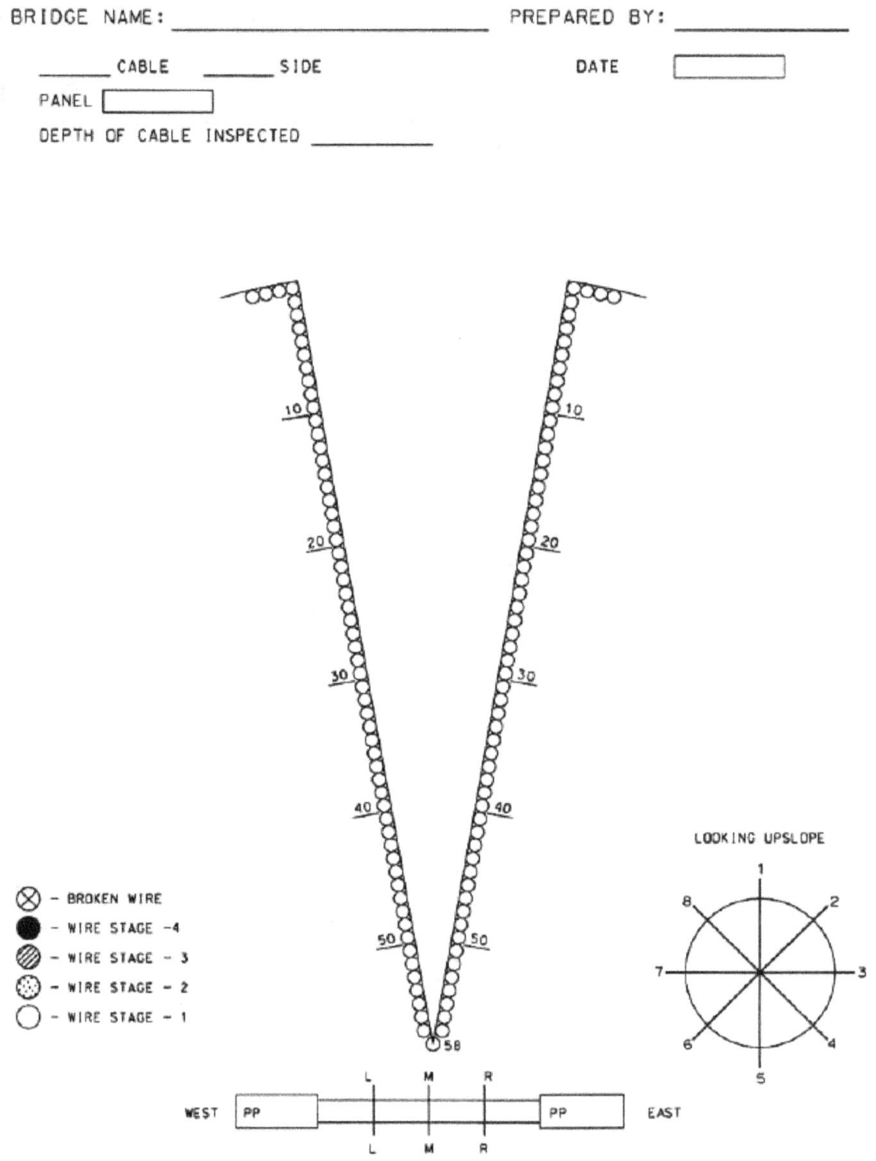

Figure 2-12 Form for recording observed wire damage inside the wedged opening (taken from *NCHRP Report 534*)

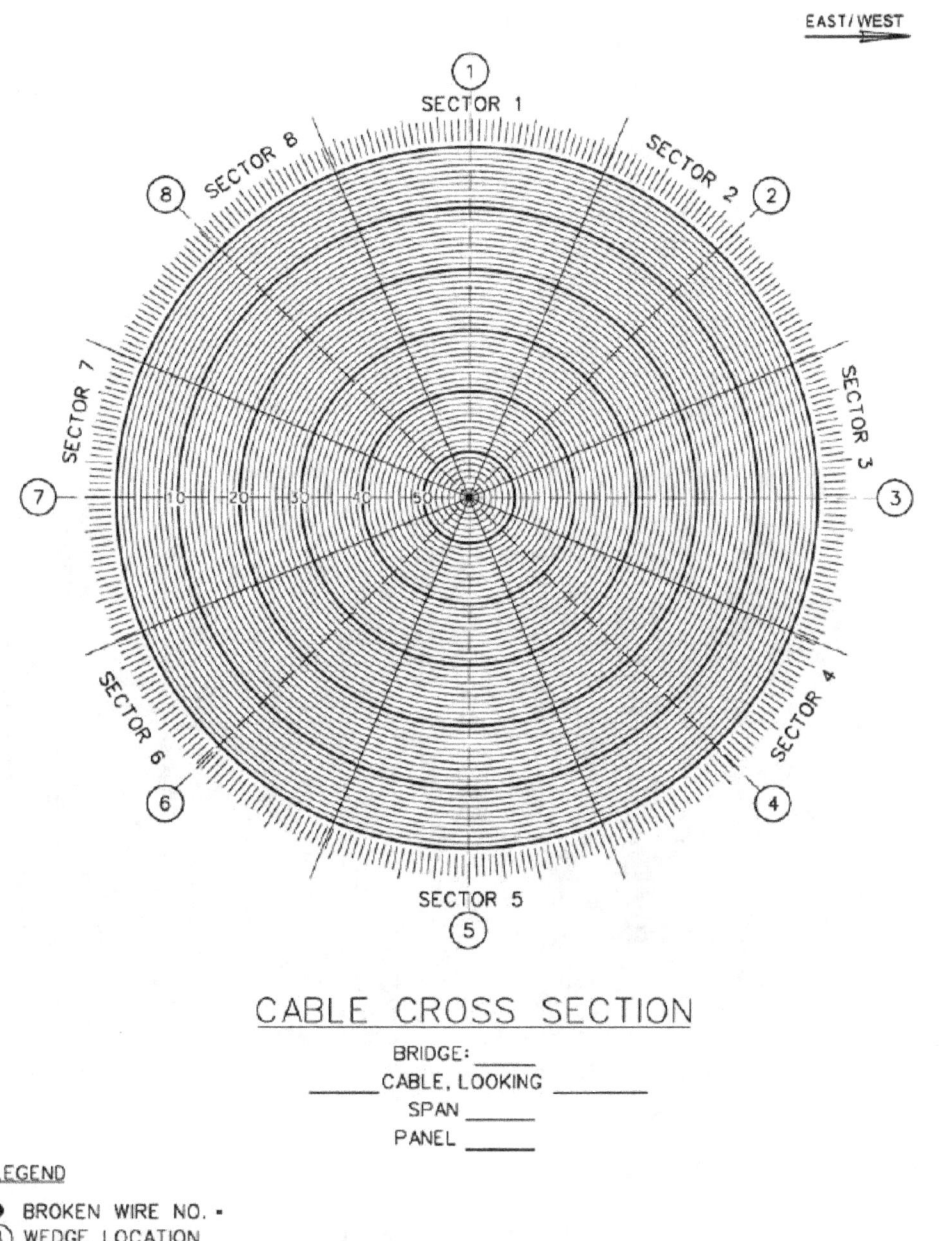

Figure 2-13 Form for recording locations of broken wires and samples for testing (taken from *NCHRP Report 534*)

- Additional Internal Inspections – The number of locations to be opened and wedged (see Figure 2-14 for typical cable wedging) after the second internal inspection depends on the conditions revealed by previous inspections, and the locations should be selected following the instructions above for second internal inspections. Specific conditions warrant additional activities:
    - Stage 4 corrosion of more than 10% of the wires/broken wires in a cable panel – The cable should be scheduled for a full interior inspection. Remedial action, such as the introduction of corrosion inhibitors, should be taken. Installation of an acoustic monitoring system is strongly recommended to detect wire breaks.
    - Stage 3 corrosion or worse found in a previous inspection – Recommended that an acoustic monitoring system be installed and monitored for a period of 12 to 18 months prior to the next internal inspection. The subsequent internal inspection locations should be selected to coincide with wire breaks, if any occur.

Figure 2-14 Photo showing a typical cable wedged open for inspection (taken from *NCHRP Report 534*)

2.1.2 Outline of Internal Inspections

The planning and mobilization for cable internal inspections are detailed in Section 2.3 of the reference document. The main items are listed below for quick cross-referencing (page number from *NCHRP Report 534*):

- General Planning and Mobilization (2-12)
- Inspection Planning (2-13)
    - Review of Available Documents (2-13)
    - Preliminary Field Observations and Cable Walk (2-14)
    - Interviews of Maintenance Personnel (2-14)

- o Inspection Forms (2-15)
- o Tool Kit (2-15)
- o Inspection QA Plan (2-17)
- o Inspection Locations (2-17)
- Construction Planning (2-17)
  - o Design of Work Platform (2-17)
  - o Construction Equipment (2-17)
    - Cable Compactors (2-17)
    - Steel Straps (2-17)
    - Wire Wrapping Machines (2-18)
    - Wedging Implements (2-18)
  - o Preparations for Suspender Removal (2-18)
  - o Replacing Wire Wrapping (2-19)
- Non-Destructive Evaluation (NDE) Techniques (2-19)
  - o Monitoring Devices (2-19)

2.1.3 Outline of Inspection and Sampling

The procedures for cable internal inspection and sampling are detailed in Section 2.4 of the reference document. The main items are listed below for quick cross-referencing (page number from *NCHRP Report 534*):

- Cable Unwrapping (2-20)
  - o Wrapping Wire Tension Tests (2-20)
  - o Removal of Wrapping Wire (2-21)
  - o Lead Paste Removal (2-21)
  - o Cable Diameter (2-21)
- Cable Wedging (2-21)
  - o Radial Wedge Locations (2-21)
  - o Wedge Initiation and Advancement (2-23)
- Wire Inspection and Sampling (2-24)
  - o Observation and Recording of Corrosion Stages (2-24)
  - o Broken Wires (2-25)
    - Wedge Spacing (2-25)
    - Wire Tracing (2-25)
    - Failed Wire Ends (2-26)
    - Sample Size (2-26)
    - Other Forms of Corrosion (2-26)
  - o Photographic Record (2-27)
  - o Measurement of Gaps at Wire Breaks (2-27)
  - o Wire Sampling (2-27)
    - Number of Samples (2-28)
    - Sample Location (2-29)
      - ➢ Stage 1 Wires (2-29)
      - ➢ Stage 2 Wires (2-29)
      - ➢ Stage 3 and Stage 4 Wires (2-30)

- - Number of Specimens in Each Sample and Length of Samples (2-30)
  - Identification of Microenvironments (2-31)
    - pH of Interstitial Water (2-31)
    - Corrosion Products (2-31)
    - Permanent Probes (2-31)
  - Cable Bands and Suspender Removals (2-31)
    - Cable Band Bolt Tension (2-31)
    - Suspender Removal and Cable Inspection (2-32)
    - Suspender Reinstallation (2-32)
  - Inspection Plan Reevaluation (2-33)
  - Reinstallation of the Cable Protection System (2-33)
  - Inspection During Cable Rehabilitation (2-33)
    - General (2-33)
    - Inspection Needs vs. Oiling Operations (2-34)
  - Inspection Testing in Anchorage Areas (2-34)
    - Wires in Strands (2-35)
    - Wires near and around Strand Shoes (2-35)
    - Eyebars (2-35)
    - Wires Inside Splay Castings (2-36)
    - Anchorage Roofs (2-36)
    - Instrumentation of Eyebars (2-36)
    - Dehumidification (2-37)
  - Inspection of Cables at Saddles (2-37)
    - Tower Saddles (2-38)
      - Tower Top Enclosures (2-38)
      - Exposed Saddles with Plate Covers (2-38)
    - Cable-Bent Saddles (2-38)
      - Saddles Inside Anchorages (2-38)
      - Extended Anchorage Housing (2-38)
      - Exposed Saddles and Plated Roofs (2-39)

2.1.4 Splicing of New Wires into the Cable

When a sample wire is removed for testing, or if a broken wire is discovered during the internal cable inspection, it is necessary to replace the removed or broken section of wire. A new portion of wire can be spliced in to the cable. However, it is generally possible to only splice wires which are within one or two inches from the surface as the deeper wires are more difficult to splice; and the smaller the cable the more important it is to splice, and it may be easier to splice. A single wire is a small fraction of the entire cable, and is redeveloped by the cable band and the wrapping wire. The typical method employed is to attach two lengths of new wire to the cut ends of the original wire with pressed-on, or swaged, ferrules. Where these two new wires meet, a threaded ferrule that acts like a turnbuckle is installed, completing the splice. In-depth details regarding the splicing of new wires into the cable are provided in Appendix D of *NCHRP Report 534*.

2.2 CABLE INSPECTION

Laboratory testing is an important part of an overall cable inspection and strength evaluation task. Test results are used to estimate wire strength, determine their stress vs. strain relationships, and ultimately evaluate cable strength. The remaining life of the wires' zinc coating can be assessed by performing additional tests.

2.2.1 Tests of Wire Properties

Strength testing is essential for evaluating cable capacity. There are five activities involved in this process:
- Specimen preparation
- Tensile tests
- Data for stress vs. strain curves
- Examining suspect wires, and
- Finding preexisting cracks

2.2.1.1 Specimen Preparation

A sample wire is defined as a length of wire removed from a cable for testing. A specimen is a piece of wire cut from the sample on which a specific test is performed. Sample wires obtained in the field should be of sufficient length to provide the number of specimens recommended in Table 2-2. All of the specimens from a given sample should be representative of the same corrosion stage.

Table 2-2 Sample lengths and number of specimens from each sample (taken from *NCHRP Report 534*)

Corrosion Stage of Sample	Minimum Number of Specimens from Each Sample			Sample Length (feet)
	Strength Tests	Weight of Zinc Tests	Preece Tests	
1	4	1	4	12
2	4	1	4	12
3	10	0	0	16 to 20
4	10	0	0	16 to 20

The cast diameter should be determined prior to cutting specimens from the sample wires.
- If the sample is of sufficient length to form a complete circle on a flat surface, measure the cast diameter in two perpendicular directions and average the results.
- If the sample is not long enough to form a complete circle, measure the rise of the arc on each of two convenient chords of the curve, calculate the resulting diameters geometrically, and average the results.

The diameter *d* is given by (equation number from *NCHRP Report 534*)

$$d = 2 \cdot \frac{(4b^2 + c^2)}{(8b)} \qquad (3.2.1\text{-}1)$$

where: b = offset between chord and arc
c = chord length

Sample wires should be inspected and assigned to the appropriate corrosion stage before the specimens are cut to suitable lengths for testing. If feasible, NDE testing (dye penetrant and magnetic flux leakage) should be performed on individual wires before they are cut, to locate preexisting cracks and ensure the worst cracks appear near the center of the specimen.

2.2.1.2 Tensile Tests

Wire strength derived from tensile tests is used to estimate cable strength. The tensile strength should be based on the nominal area of the wire. Tensile tests should be performed in accordance with ASTM A586 and ASTM A370 to determine the following wire properties:
- Breaking load in the wire
- Yield strength (0.2% offset method)
- Tensile strength
- Elongation in 10-inch-gage length
- Reduction of area
- Modulus of elasticity

2.2.1.3 Obtaining Data for Stress vs. Strain Curves

In addition to the tests listed above, wire elongation should be recorded at intervals of tensile force up to the maximum force preceding failure. The data should be used to construct a full stress vs. strain curve for each specimen. The ultimate strain corresponding to tensile strength should also be determined.

2.2.1.4 Fractographic Examination of Suspect Wires

The fracture surface of the wires should be observed using a stereoscopic optical (light) microscope and/or a scanning electron microscope to detect whether failure is ductile or brittle. A ductile failure of the wire is indicated by necking, or the reduction of the wire diameter at failure. A brittle failure exhibits pitting or cracking, failure soon after the yield point is reached, a reduction in elongation and strength, and little or no reduction in cross-sectional area.

Any fracture surface that exhibits traces of corrosion or contamination should have an X-ray energy dispersion spectral analysis performed. In addition, enlarged images of failure morphologies should be interpreted by metallurgists or corrosion experts. The images may indicate embrittlement, hydrogen-assisted cracking or other corrosion mechanisms, which the experts can identify.

2.2.1.5 Examination of Fracture Surface for Pre-existing Cracks

Cracked wires are treated as a separate group in estimating cable strength. Pre-existing cracks are defined as cracks that are present in the specimen prior to testing. They are discovered by examining the fracture surface of all tension specimens under a stereoscopic optical (light) microscope at 20X magnification. A sample wire is considered to contain a crack if any of the specimens cut from the sample contains a pre-existing crack.

A cracked specimen should be photographed (see Figure 2-15) and measured. The depth of crack and wire diameter at the failure plane should be reported in both absolute terms and as a fraction of wire diameter. In the vicinity of a brittle fracture:
- The outer surfaces of the wire should be examined under a stereoscopic optical microscope for additional pre-existing cracks.
- Longitudinal sections of short wire segments should be examined under either an optical or scanning electron microscope.

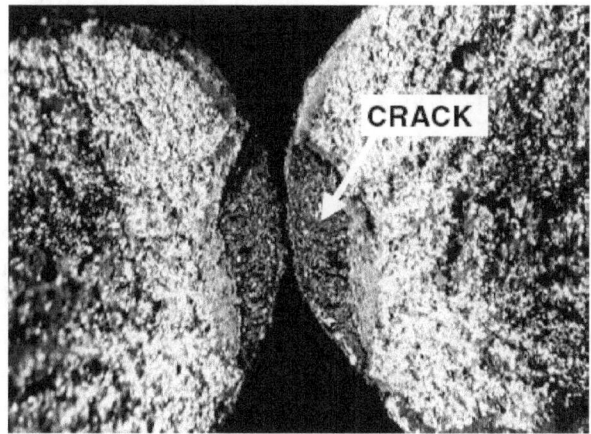

Figure 2-15 Photograph of a cracked wire (taken from *NCHRP Report 534*)

2.2.2 Zinc Coating Tests

Two types of tests are performed on the zinc coating during cable wire evaluation:
- Weight of zinc tests
- Preece

Both of these tests should be conducted on Stage 1 and Stage 2 specimens that exhibit uniform zinc or spotty zinc loss. The minimum depth of the coating determines its condition, not the average depth.

2.2.2.1 Weight of Zinc Coating

The Weight of Zinc Coating Test, specified in ASTM A90, is a gravimetric test that measures the weight of zinc removed from a unit length of wire. It is used to determine the average weight of zinc in that length, separate from variations in coating thickness.

The average weight of zinc in a unit length, determined by testing, can be converted to an average remaining thickness of zinc coating and used to estimate when the zinc coating will be depleted.

2.2.2.2 Preece Test for Uniformity

The Preece Test, specified in ASTM A239, is used to determine the uniformity of the zinc coating on Stage 1 and Stage 2 wires. It is a chemical test that depends on the reaction of copper sulfate and zinc. It is used to confirm whether the coating on the specimen is depleted uniformly or locally. The Preece Test is more important than the Weigh of Zinc Coating Test, because it is a better indicator of when the zinc is depleted, since only a small depletion in the zinc coating is needed for the onset of pitting and cracking.

Preece Tests are performed in series terminated after the fourth dip. Wires are dipped in a copper sulfate solution for a standard time period. If sufficient zinc is present, then the wire retains its shiny surface from the intact zinc. If the zinc is insufficient, then the copper electroplates the steel, and the wire surface turns the color of copper.

2.2.3 Chemical Analysis

The chemical composition of the steel wire should be determined under any of these circumstances: tests were never performed, results from previous tests are unavailable, or tests reveal unusual variations in the tensile strength of samples. Percentages of the following elements should be obtained:
- Carbon
- Silicon
- Manganese
- Phosphorous
- Sulfur
- Copper
- Nickel
- Chromium
- Aluminum

A minimum of five wires should be analyzed for completeness. If the steel's chemistry varies significantly, a metallurgist should be consulted to determine the effects on the wire's properties.

A chemical analysis of the surface deposits (corrosion present) on the wire samples should be performed to detect harmful contaminants. The results should be reported in absolute amounts, per unit of wire area. Determine the presence or absence of the following salts:

- Chloride
- Sulfates
- Nitrates

2.2.4 Corrosion Analysis

An investigator may recommend studying the corrosion product on a wire or anchorage. Corrosion analyses are typically performed on surface corrosion films, the fracture surfaces of steel, or corrosion by-products. Chlorides from roadway salts, as well as sulfates and nitrates from acid rain, are associated with causing corrosion.

The following types of electronic microscopy are used in corrosion analysis:
- X-Ray Photoelectron Spectroscopy (ESCA)
- Energy Dispersive X-Ray Analysis (EDAX)
- X-Ray Diffraction (XRD)

3.0 EVALUATION OF FIELD AND LABORATORY DATA OVERVIEW

This section provides an overview of the tabulated field observations, and using these field observations for estimating cable strength. Items presented in this section include:
- Mapping and estimating wire deterioration
- Wire strength properties based on testing
- Wire force redevelopment

The above items are discussed in detail in the *NCHRP Report 534*; therefore, these topics are only highlighted in this section. The reader is encouraged to review *NCHRP Report 534* for detailed discussions. Many of the calculations provided herein, as they pertain to the wire deterioration, strength properties, and estimation of cable strength, lend themselves ideally to the use of spreadsheets.

3.1 MAPPING AND ESTIMATING WIRE DETERIORATION

The aim of mapping wire deterioration is to establish the mean size of the population of each of the damage stages (visually) and of the cracked stage which is a subdivision of Stage 4. Broken wires are also separate populations. The sum of all populations shall amount to the total number of wires in the cable.

Mapping and the wire deterioration estimation in the cable cross-section is developed directly from the field inspection data. For each side of the wedged opening, an observed wire on the side of the wedged opening is assumed to represent all wires at the same depth for that half sector.

It is assumed that the cable is composed of concentric rings of wires, arranged around one central wire. This assumption facilitates the recording and analyzing data gathered in the field. It should be noted however, that the wires in the cable do not actually lie in precise rings around the center of the cable.

The number of rings and the subsequent number of wires in each ring should be determined prior to inspection. This allows the inspectors to develop the necessary documentation for recording wire deterioration prior to the actual inspection.

3.1.1 Number of Rings in the Cable

The number of concentric rings in the cable, not including the center wire, is estimated by:

$$X = \sqrt{\frac{N}{\pi}} + 0.5 \text{, rounded to the next highest integer} \qquad (4.3.1.1\text{-}1)$$

where: X = number of rings in the cable, not including center wire
        N = actual number of wires in the cable

Once the number of rings in the cable has been determined, figures such as Figure 2-12 and Figure 2-13 (Section 2.1.1.3) can be developed for recording purposes during the actual inspection.

3.1.2 Number of Wires in Each Ring

The number of wires in each concentric ring is determined by:

$$n_x = \frac{2x(N-1)}{X(X+1)} \qquad (4.3.1.2\text{-}1)$$

where: n_x = number of wires in ring x
 x = number of rings from the center of the cable to the specific ring being considered

3.1.3 Wire Deterioration/Corrosion Mapping

The deterioration of the wires will typically be recorded during the inspection on a form similar to Figure 3-1. Data such as this is collected for each wedged opening; typically, eight wedge openings are used. For each side of the wedged opening, an observed wire on the side of the wedged opening is assumed to represent all wires at the same depth for that half sector, resulting in Figure 3-2. The wires in each half sector are assigned the corrosion stage of the observed wire on the side of the wedged opening. The corrosion map provides a visual representation of the amount of corrosion in the cable, allowing the engineer to better understand what was observed in the field during the actual inspection.

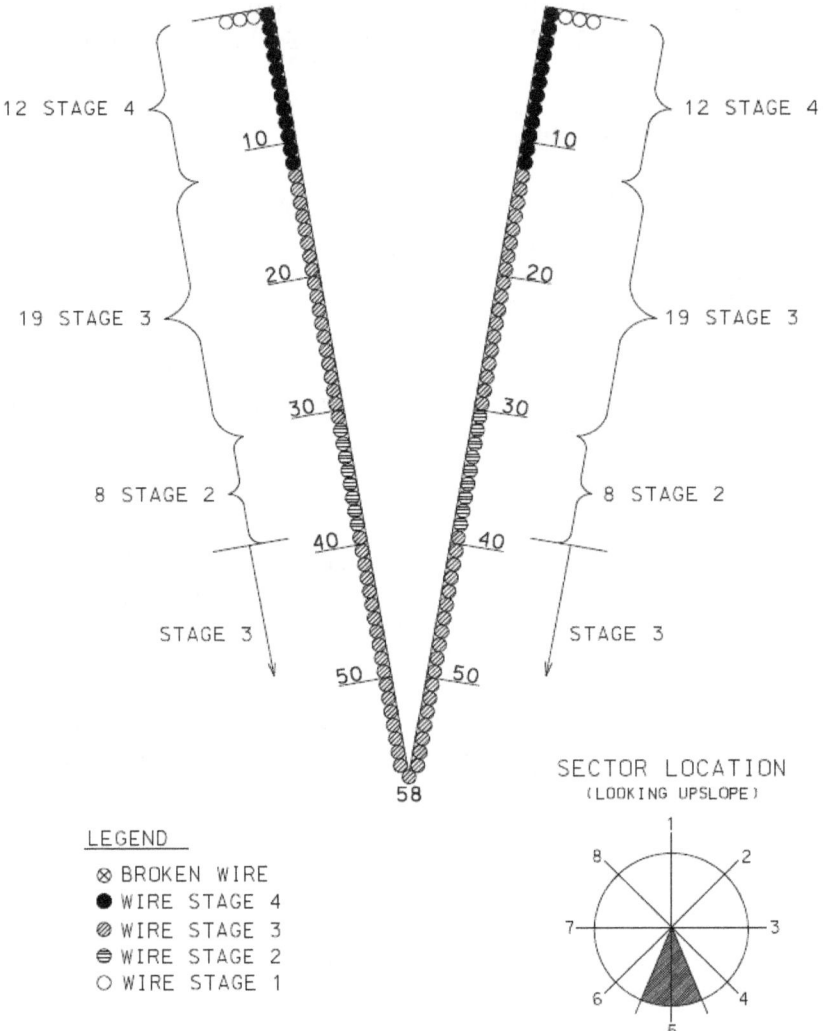

Figure 3-1 Typical form for recording observed wire deterioration

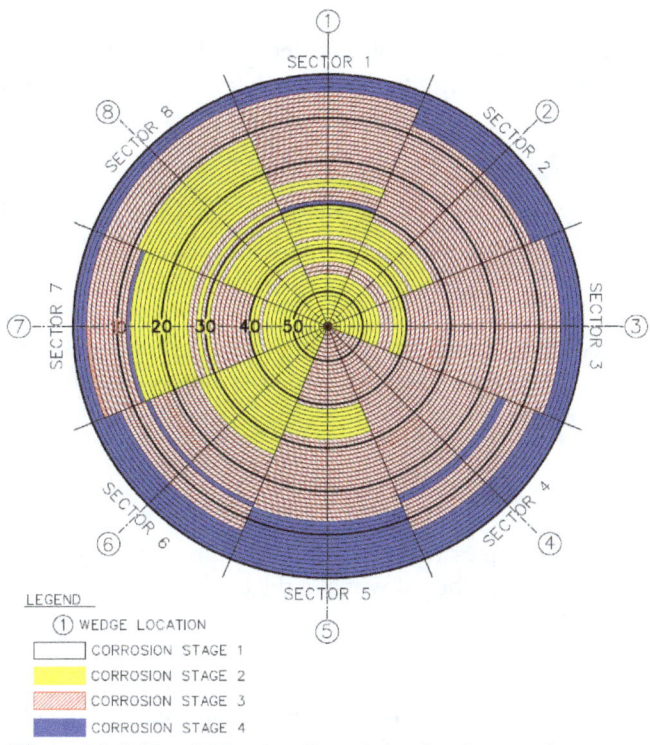

Figure 3-2 Typical wire deterioration/corrosion map

3.1.4 Fraction of Cable in Each Corrosion Stage

As shown in Figure 3-1, each observed wire is assigned a corrosion stage. The assigned corrosion stage is based on the visual inspection of the wire, as described in section 1 of this document and Figure 1-9. The total number of wires in each stage of corrosion, k, is calculated by adding together the individual wires in each half-sector.

The number of wires in a given half-sector for each stage of corrosion can be determined from the equation:

$$N_{jk} = n_{jk} \cdot a_{jk} \qquad (4.3.2\text{-}3)$$

where: k = corrosion stage of wires (k = 1, 2, 3, and 4)
jk = identification number of an observed wire in Stage k
N_{jk} = number of wires in half-sector represented by observed wire jk
a_{jk} = fraction of a circle corresponding to the width of the half-sector that contains observed wire jk. For a cable inspected with four wedge lines (quadrants), the fraction is 1/8; when eight wedge lines are used, the fraction is 1/16. Whenever all half-sectors are of equal size, a_{jk} is a constant
n_{jk} = total number of wires in ring that contains observed wire jk

The total number of Stage *k* wires in the cable is calculated by adding the Stage *k* wires from each ring in the cable. In equation form, the total number of wires in each stage (N_{sk}) is:

$$N_{sk} = \sum_{jk=1}^{J_k} N_{jk} = \sum_{jk=1}^{J_k} n_{jk} \cdot a_{jk} \qquad (4.3.2\text{-}4)$$

where: J_k = total number of observed wire sin Stage *k*
N_{sk} = number of Stage *k* wires in the cable

As a check, the sum of all the N_{sk} for all stage should equal the total number of wires (*N*) in the cable. The fraction of the wires represented by Stage *k* is simply calculated as:

$$p_{sk} = \frac{N_{jk}}{N} \qquad (4.3.2\text{-}5)$$

where: p_{sk} = fraction of wires in the cable represented by Stage *k*
N = total number of wires in the cable

3.1.5 Number of Broken Wires

The method of estimating the number of broken wires in an inspected panel depends on the location of the broken wires. In the wedged openings, if broken wires are found beyond the first few rings, each broken wire at the given depth in the cable is assumed to represent all wires in that particular ring for that half-sector of the wedged opening. The method of estimating the number of wires in the cable interior is discussed in Article 4.3.3.1 of *NCHRP Report 534*.

When broken wires are found only at the outer surface, or in the first few rings from the outside of the cable, the number of broken wires in the cable should be estimated in accordance with Article 4.3.3.2 of *NCHRP Report 534*, as further discussed within this section.

When broken wires are found mostly in the outer ring of the cable, the depth at which broken wires are no longer found, d_o, can be determined from observing the wedged openings. The number, location, and depth of the broken wires should be recorded during the cable inspection, using Figure 2-13. Additional wedged openings, with wedges driven at least 2 inches beyond the depth of corrosion Stage 4 wires, may be required to ensure the depth of the broken wires.

The depth, d_o, is expressed as the number of rings from the cable outer surface, with the outer ring being assigned the number 1. The number of broken wires in each ring is conservatively assumed to decrease in a linear manner from the outer ring to zero broken wires at the depth d_o. In this case, the total number of broken wires in the cable cross section of the inspected panel is approximated by:

$$n_{bi} = n_{b1,i} \cdot \frac{d_o}{2} \qquad (4.3.3.2\text{-}1)$$

where: n_{bi} = total number of broken wires in cable cross section for given panel
$n_{b1,i}$ = number of broken wires in the outer ring of the cable in panel i
d_o = depth into the cable at which no broken wires are found

3.2 WIRE PROPERTIES

The properties of a single cable wire can vary from those of other wires in the cross section, and also vary along the length of the same wire. The lowest values of these properties in a panel length should be determined on a probabilistic basis for all populations of wires of given corrosion stage..

The first step is to assign the sample wires into sorted groups based on the laboratory testing results. These groups are slightly different from the corrosion stages previously specified. These groups are:
- Group 1 - samples exhibiting Stage 1 corrosion, if determined by the investigator to be needed
- Group 2 - samples exhibiting Stage 1 or Stage 2 corrosion
- Group 3 - samples exhibiting Stage 3 corrosion that are not cracked
- Group 4 - samples exhibiting Stage 4 corrosion that are not cracked
- Group 5 - samples exhibiting Stage 3 or Stage 4 corrosion that contain one or more cracks

Generally, the properties for Stage 1 and Stage 2 wires vary so little that they can be considered in a single group, Group 2. However, if there are significant variations between corrosion Stage 1 and Stage 2, then the Group 1 category should be used accordingly.

Each wire sample taken from the cable is divided in to 11 specimens. As discussed in the section addressing the laboratory investigation of these specimens, tensile tests will be conducted. The mean and standard deviation of the tensile strengths and the ultimate elongation of each specimen cut from each wire sample can then be determined based on the laboratory testing. With these mean individual wire properties, in conjunction with the groupings above, the minimum tensile strength of each group can be estimated.

3.2.1 Cracked Wires as a Separate Group

A sample wire is considered to be cracked if any of the tested specimens cut from the sample contains a pre-existing crack. The fraction of cracked wires in each stage of corrosion is given simply by:

$$p_{c,k} = \frac{\text{number of cracked Stage k sample wires}}{\text{total number of Stage k sample wires}} \qquad (4.4.2\text{-}1)$$

where: $p_{c,k}$ = fraction of Stage k wires that are cracked

However, for Stage 3 wires in particular, a 0.33 factor is included to adjust for the fact that Stage 3 wires found deeper inside the cable rarely exhibit cracks. However, if cracks in Stage 3 wires are found deeper in the cable, the factor should be increased accordingly. Therefore, for stage 3 wires only:

$$p_{c,3} = \frac{0.33 \cdot \text{number of cracked Stage k sample wires}}{\text{total number of Stage k sample wires}} \qquad (4.4.2\text{-}2)$$

where: $p_{c,3}$ = fraction of Stage 3 wires that are cracked

Due to the difficulty in removing wires that are deep inside the cable, most Stage 3 samples are taken from areas of not more than 15 wires (approximately 3 inches) from the cable surface. Therefore, the fraction of stage 3 wires that are cracked is more than likely biased because all of the cracks in Stage 3 wires are generally found near the interface between Stage 3 and Stage 4 wires. To estimate this, several curves of the fraction of cracked wires versus the depth into the cable were developed for a given bridge. This analysis indicated that in estimating the cable strength, the average number of Stage 3 wires in the cable that are cracked should be assumed to be about one-third (0.33) of the number of Stage 3 wires that testing revealed to be cracked [5]. Again, this factor should be adjusted if cracks in Stage 3 wires are found deeper in the cable, and can be increased to a value as much as 1.00 if required.

3.2.2 Individual Wire Properties

3.2.2.1 Mean Properties

Each wire sample taken from the cable is divided in to 11 specimens, all having a length of 12 inches. As discussed in the section addressing the laboratory investigation of these specimens, a tensile test will be conducted. The sample mean and sample standard deviation, as well as the estimated minimum strength properties of the wire in the panel, can be estimated based on this laboratory testing. The mean and standard deviation for each wire sample is calculated using typical statistical equations such as:

$$\mu_{sj} = \frac{1}{n_j} \sum_{i=1}^{n_j} x_{i,j} \qquad (4.4.3.1\text{-}1)$$

$$\sigma_{sj} = \sqrt{\left(\frac{1}{(n_j - 1)} \sum_{i=1}^{n_j} x_{ij}^2 \right) - \mu_{sj}^2} \qquad (4.4.3.1\text{-}2)$$

where:
j = number identification of wire sample
i = number identification of specimen (portion of wire that is a part of sample j)
μ_{sj} = sample mean of the property x for sample j
σ_{sj} = standard deviation of the property x for sample j
x_{ij} = property of specimen i cut from sample j
n_j = number of specimens tested from sample j

3.2.2.2 Minimum Properties in a Panel Length

The weakest point of the wire within the panel must be estimated. The probable minimum tensile strength of each wire sample within in the panel length is given by:

$$x_{1,j} = \mu_{sj} + \Phi^{-1}\left(\frac{L_0}{L}\right) \cdot \sigma_{sj} \tag{4.4.3.2-1}$$

where: $x_{1,j}$ = probable minimum value of x_j in a length L of the wire from which sample j is removed
μ_{sj} = sample mean of the property x for sample j (see Eq. 4.4.3.1-1)
σ_{sj} = standard deviation of the property x for sample j (see Eq. 4.4.3.1-2)
L_0 = length of the test specimen between grips of the testing machine (specimens for tensile test should measure 12 inches between the grips of the testing machine)
L = length of a wire between centers of cable bands (panel length)
$\Phi^{-1}(L_0/L)$ = inverse of the standard normal cumulative distribution for the probability L_0/L

The value of $\Phi^{-1}\left(\frac{L_0}{L}\right)$ is determined using Figure 3-3 (Figure 4.4.3.2-1 of *NCHRP Report 534*).

If a negative value of $x_{1,j}$ results, zero should be used as the minimum value.

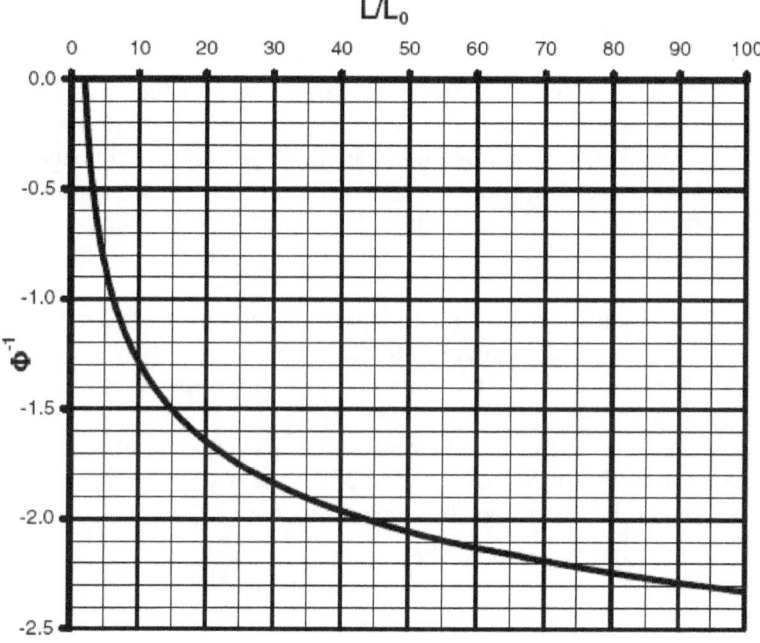

Figure 3-3 Graph for computing the inverse of the standard normal cumulative distribution function (taken from *NCHRP Report 534*)

3.2.3 Wire Group Mean Strength and Standard Deviation

Once the individual wire properties are determined, it is then possible to determine the mean properties (tensile strength, ultimate strength, etc.) for each corrosion stage group of wires. For each group listed in section 3.2, the sample mean and standard deviation of the minimum estimated properties are calculated using:

$$\mu_{sk} = \frac{1}{n_k} \sum_{i=1}^{n_k} x_{i,j} \qquad (4.4.4\text{-}1)$$

$$\sigma_{sk} = \sqrt{\left(\frac{1}{(n_k - 1)} \sum_{i=1}^{n_k} x_{i,j}^2\right) - \mu_{sk}^2} \qquad (4.4.4\text{-}2)$$

where:
k = corrosion stage group identification of the wires (typically, k = 2, 3, 4, or 5)
j = number identification of wire sample
μ_{sk} = sample mean of the property x for Group k
σ_{sk} = standard deviation of the property x for Group k
$x_{1,j}$ = probable minimum value of x_j in a length L of the wire from which sample j is removed
n_k = number of specimens tested from Group k

3.3 WIRE REDEVELOPMENT

A broken wire does not become inactive over the entire length of cable. The broken wire will redevelop its force as the distance from the break increases. This increase in wire force is caused by the friction at the cable bands. The force in a wire developed at a cable band is estimated using the measured gap between the ends of the broken wire. Also, the gap that develops when a wire is cut for the purpose of removing a sample can be used for estimating the force developed in the wire.

During the cable inspection, investigators should take measurements of as many failed wire separations as possible in the panel being investigated. Measurements from gaps that result from the taking of wires samples should also be recorded by the investigators.

Further discussion with regard to the calculations for the development of wire forces at cable bands is provided in Article 4.5 of *NCHRP Report 534*. The reader should refer to this article if the force available from broken wires needs to be considered.

4.0 ESTIMATION OF CABLE STRENGTH

Several factors contribute to the estimated strength of the cable. These factors include:
- The tensile strength of the wires.
- The distribution of the tensile strength.
- The broken wires in the cable.
- The cracked wires in the cable.
- The redevelopment force in broken wires due to friction.

The tensile strength of the wires and its distribution are determined by assigning the exposed wires to the various stages of corrosion (Section 1.3.4) and counting them, and from testing sample wires removed from the cable. Broken wires are counted during the inspection. The number of cracked wires is determined by tensile tests of the sample wires. The ability of friction to redevelop forces in the wires is estimated by measuring the gaps between the ends of broken or cut wires inside the cables.

The estimated strength is calculated at a specific inspected location along the cable, referred to herein as the evaluated panel. The estimated strength of the cable for the evaluated panel can be taken as the sum of the following categories:
- All wires in the evaluated panel minus broken wires in the evaluated panel as well as nearby panels
- Wires that are cracked in nearby panels, affecting the strength of the same wires in the evaluated panel
- Wires that are broken in nearby panels, affecting the strength of the same wires in the evaluated panel based upon effective redevelopment length.

Methods for estimating the strength of unbroken wires are presented within this section. Wires that are broken in panels adjacent to the evaluated panel share in the cable tension because of friction that develops at the cable bands. The redevelopment of broken wires in panels adjacent to the evaluated panel is not provided within this section of this Primer, but can be found in Article 5.3.4 of the *NCHRP Report 534*.

The strength of unbroken wires can be determined by one of three methods:
- Simplified Model
- Brittle-Wire Model
- Limited Ductility Model

The Simplified Model could be applied to cables that have very few cracked wires, with an upper limit of 10% of the total wire population. If the 10% limit is exceeded, the Brittle-Wire Model should be used. The Simplified Model is based on the Brittle-Wire Model, but subtracts all cracked and broken wires, and uses a single distribution curve for the tensile strength of the unbroken wires. The Brittle-Wire Model assumes that all the wires are subjected to the same tensile stress at any given strain. The Limited Ductility Model should be used if the wires display unusual variations in tensile strength, which would be reflected in stress-strain curves developed from the laboratory testing. In the Limited Ductility Model, the ultimate strain of the wires is used as the variable in the distribution functions.

4.1 WIRE GROUPINGS

As discussed previously, the wires are assigned to various groups based on the stage of corrosion, or if they are cracked. The cracked wires are separated into Group 5, regardless of the corrosion stage. The number of cracked wires in the effective development length is taken as N_5. The estimated number of broken wires is treated separately from cracked wires.

For the treatment of broken or cracked wires, the reader should refer to Article 5.3.2 of *NCHRP Report 534*. In this Article, equations are provided for calculating:
- Broken Wires in the Effective Development Length (5.3.2.1)
- Repaired Wires in the Effective Development Length (5.3.2.2)
- Unbroken Wires in Each Stage of Corrosion (5.3.2.3)
- Discrete Cracked Wires in the Effective Development Length (5.3.2.4.1)
- Redevelopment of Cracked Wires That Fail (5.3.2.4.2)
- Effective Number of Unbroken Wires (5.3.2.5)

Many of the calculations within Article 5.3.2 lend themselves to the use of spreadsheets given the enormous amount of data that can result.

4.2 STRENGTH OF UNBROKEN WIRES

The strength of the cable with unbroken wires can be estimated using one of three strength models, which vary in the amount of calculations required. The Limited Ductility model is the most rigorous; while the Brittle-Wire Model and the Simplified Model employ simplifying assumptions.

The Limited Ductility Model and Brittle-Wire Models are used to estimate the strength of a cable composed of many wires that are subjected to the same strain. The Simplified Model subtracts all cracked and broken wires and uses a single distribution curve for the tensile strength of the remaining unbroken, uncracked wires; which leads to a conservative estimate of the cable strength.

This Primer will provide full details for the Simplified Strength Model, and provide preliminary details of the Brittle-Wire and Limited Ductility Strength Models. For in-depth details of the Brittle-Wire and Limited Ductility Models, the reader should refer to Articles 5.3.3.2 and 5.3.3.3, respectively, of *NCHRP Report 534*.

4.2.1 Simplified Strength Model

The Simplified Strength Model is a simplification of the Brittle-Wire Strength Model discussed below. Cracked and broken wires are assumed to not contribute to the estimated strength of the cable. Therefore, the Simplified Strength Model should be used when the cable is found to have very few cracked wires. To use the Simplified Model, no more than 10% of the entire population of wires should be cracked. If more than 10% of the wires are cracked, the investigator should consider the use of the Brittle-Wire Strength Model.

In most cases, the Simplified Strength Model will underestimate the cable strength. However, the Simplified Strength Model is useful in locating the most severely deteriorated panel among the panels evaluated. Once the "worst-case" panel is identified, a more complex strength models, such as the Brittle-Wire, can be employed to obtain a reasonable estimate of the cable strength.

4.2.1.1 Mean Tensile Strength of Uncracked Wires

The estimated mean strength (μ_s) of the entire cable is calculated by taking the fraction of the cable represented by Groups (k) 2, 3, and 4 combined with their respective sample mean values of minimum tensile strength. Similarly, the standard deviation (σ_s) of the tensile strength of the combined groups can be calculated. The following calculations should be used for each:

$$\mu_s = \sum_{k=1}^{4}(p_{uk} \cdot \mu_{sk}) \qquad (5.3.3.1.1\text{-}1)$$

$$\sigma_s = \sqrt{\left(\sum_{k=2}^{4} p_{uk}(\sigma_{sk}^2 + \mu_{sk}^2)\right) - \mu_s^2} \qquad (5.3.3.1.1\text{-}2)$$

in which:
$$p_{uk} = \frac{N_k}{N_{eff} - N_5} \qquad (5.3.3.1.1\text{-}3)$$

where:
- k = group identification of the wires (typically, k = 2, 3, 4, or 5)
- μ_s = sample mean tensile strength of the combined group of wires (entire cable cross-section), excluding cracked wires
- μ_{sk} = sample mean tensile strength of Group k
- σ_s = sample standard deviation of the tensile strength of the combined group of wires (entire cable cross-section), excluding cracked wires
- p_{uk} = fraction of unbroken and uncracked wires in the cable represented by Group k
- N_{eff} = effective number of unbroken wires in the evaluated panel (see Article 5.3.2.5 of *NCHRP Report 534*)
- N_5 = number of discrete cracked wires in the effective development length (see Article 5.3.2.5 of *NCHRP Report 534*)
- N_k = number of Group k wires in the evaluated panel

4.2.1.2 Cable Strength Using the Simplified Model

The cable strength is calculated from the equation:

$$R = (N_{eff} - N_5) \cdot a_w \cdot \mu_s \cdot K \qquad (5.3.3.1.2\text{-}1)$$

where: K = reduction factor as a function of the coefficient of variation, σ_s/μ_s (given in Figure 5.3.3.1.2-1 of *NCHRP Report 534*), see Figure 4-1 of this Primer
R = estimated cable strength
a_w = nominal area of one wire used in the lab analysis.

The strength reduction factor K, is the Brittle-Wire Strength of the combined groups of uncracked wires, divided by the product of the mean tensile strength of the combined groups of uncracked wires and total area of uncracked wires. The derivation of the reduction factor, *K*, is given in Appendix A of *NCHRP Report 534*. The factor K, is based on the fact that a compound strength distribution used in the Brittle-Wire model is replaced by a single Weibull strength distribution for the Simplified Model

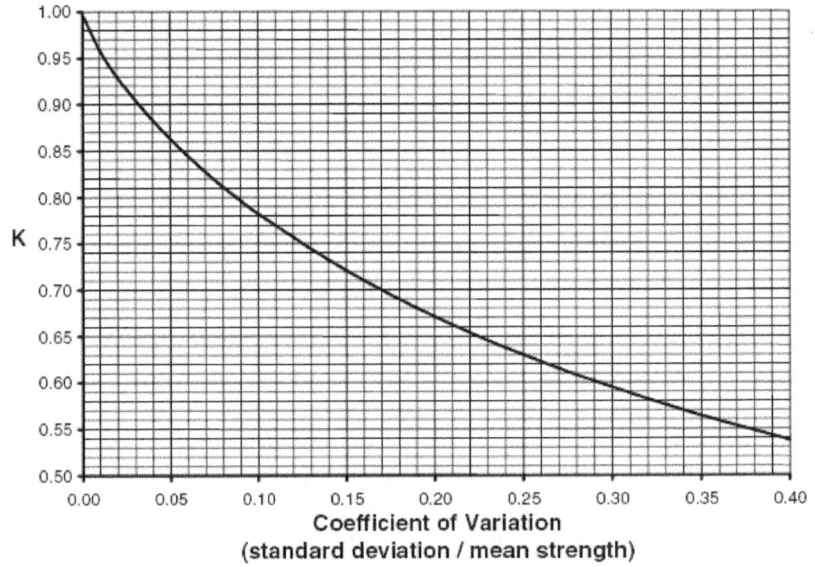

Figure 4-1 Graph for computing the strength reduction factor, *K* (taken from Figure 5.3.3.1.2-1 of *NCHRP Report 534*)

4.2.2 Brittle-Wire Strength Model

Per *NCHRP Report 534*, the Brittle-Wire Strength Model is the recommended model for determining the estimated strength of the cable. The Brittle-Wire Strength Model assumes that all of the wires follow the same stress-strain diagram. Therefore, it is assumed that each wire will bear an equal portion of the total load and each wire will have the same stress. The model is a simplification of the Limited Ductility model, because the Brittle-Wire Model assumes that a wire fails immediately upon reaching its ultimate stress. The load carried by the broken wire is then assumed to transfer equally to the remaining unbroken wires.

The Brittle-Wire Model is somewhat conservative because it ignores the ability of the individual wires to behave plastically. The unbroken wires will not break at the individual ultimate stress, but will actually deform plastically until fracture, allowing additional wire strength that is unaccounted for in the Brittle Wire Model.

The cable strength is calculated through an iterative process of incrementally increasing the cable stress and determining the number of wires that fail at each load increment. Each wire is assumed to fail when it reaches its tensile strength, as determined through laboratory testing. The number of failed wires is subtracted from the number of previously intact wires to determine the new number of unbroken wires. The cable force is calculated by taking the area of the unbroken wires at the given increment, multiplied by the stress in the wire. The process is repeated until the maximum force in the cable is obtained; at a certain level of stress the wires will fail faster than the wire force can be increased. This process is represented by what is known as a survivor function, or reliability function.

The cable strength can be estimated by using a sorted list of the wires in order of tensile strength, or by using a statistical distribution of the tensile strength such as a Weibull cumulative distribution function. The fraction of the cable represented by each group, p_k, and the Weibull distribution curves for tensile strength of the specimens representing each of the groups can be combined to determine the compound distribution curve for the entire unbroken wire population.

The reader is referred to Article 5.3.3.2 of *NCHRP Report 534* regarding the implementation of the Brittle-Wire Strength Model. Appendix Article A.5 provides information regarding the calculation of parameters for estimating the cable strength using the Weibull distribution. Furthermore, there are two examples in Appendix C that demonstrate the use of the Brittle-Wire Model, with and without considering different deterioration effects in adjacent panels.

4.2.3 Limited Ductility Strength Model

The Limited Ductility Strength Model should be used if the wires display an unusual variation in tensile strength, which would be reflected in the stress-strain curves. However, the Limited Ductility Strength Model is quite complex, and should be used only if it is deemed absolutely necessary by the bridge engineer. In the Limited Ductility Model, a wire is assumed to fail when it elongates enough to reach its individual ultimate strain.

This model requires the determination of the ultimate strain of each wire specimen through laboratory testing, and developing a stress-strain diagram for each wire sample. For a specific value of strain, the unbroken wire is subject to a tensile stress that corresponds to the stress-strain diagram for that particular wire. A wire is assumed to fail when it reaches its maximum elongation, at which time the load carried by the subject wire is transferred equally to remaining unbroken wires.

The cable strength is calculated through an iterative process, similar to that discussed for the Brittle-Wire Model, but with strain values instead of stress. The iterative process is repeated until the maximum force in the cable is obtained, as at some threshold point, wires will fail faster than the cable force can be increased. The cable strength can be estimated by using a sorted list of the wires in order of ultimate elongation, or by using a statistical distribution of the ultimate elongation, such as a Weibull cumulative distribution function.

The reader is referred to Article 5.3.3.3 of *NCHRP Report 534* regarding the implementation of the Limited Ductility Strength Model. Appendix Article A.5 provides information regarding the calculation of parameters for estimating the cable strength using the Weibull distribution.

4.3 NON-APPLICABILITY OF LOAD AND RESISTANCE FACTOR RATING (LRFR)

The AASHTO *Manual for Bridge Evaluation* [6] provides a methodology for load rating a bridge consistent with the load and resistance factor design philosophy of the AASHTO *LRFD Bridge Design Specifications* [7]. Load and Resistance Factor Rating (LRFR) is a reliability-based evaluation methodology where the load and resistance factors have been calibrated to achieve a uniform and consistent level of reliability. This entails the consideration of applied load effects, in combination with the known member resistances, to verify that the target failure probabilities are not exceeded.

The LRFR calibration, as with the LRFD calibration, are geared to safety targets considered appropriate for short to medium span bridges. The LRFR procedures adopt a target reliability index of approximately 2.5, which has been calibrated to past AASHTO operating level load ratings. This is a reduced value from the LRFD reliability index of 3.5, which was derived for a severe traffic-loading case. The reliability index of 2.5 was chosen for LRFR so as to reflect the reduced exposure period, consideration of site realities, and the economic considerations of rating versus design (section C6A.1.3 [6]). However, for long-span bridges, such as a suspension bridge, no similar calibration has been performed for LRFR, and therefore the current LRFR is not directly applicable to the rating of suspension bridge cables.

For long span bridges, a more conservative safety target may be appropriate due to the consequences in the event of failure. Additionally, the current LRFR calibration is meant for bridges with dead load to live load ratios that are generally not higher than about 2:1. In long-span bridges the dead loads may be higher than the live loads by a factor of 5 or higher. Such ratios will exceed the calibration assumptions of LRFR. Further research is required in the area of LRFD and LRFR, before specific recommendations regarding load factors and rating strategies can be made for long span suspension bridges. The bridge owner may also specify criteria related to load factors to be used for a specific suspension bridge rating.

5.0 INSPECTION DOCUMENTATION, REPORTING, AND RECOMMENDATIONS

Following the completion of each suspension bridge parallel wire cable inspection, a report is required, which collectively, over time, forms a written historical record of the cable's condition, and helps the bridge owner make informed decisions about maintenance schedules and budgets.

5.1 MAINTENANCE PERSONNEL INSPECTION REPORTS

A written report should be prepared for each periodic cable inspection performed by bridge maintenance personnel. The report should include (at minimum) the following:

- Date of Inspection
- Weather and Temperature
- Portion of Cable Inspected (e.g. west main span, south anchorage, tower saddles and cable housings, etc.)
- List of Deficiencies (identified by panel number)
- One-page summary of each deficiency including:
    o Verbal description (e.g. peeling paint, rust stains, broken wrapping, etc.)
    o Color photographs
    o Recommended action items
- Summary List of Recommended Action Items (Priority Order)

A follow-up report should be prepared for each maintenance action performed by bridge maintenance personnel. The report should include the following:

- Description of action item and maintenance performed
- Representative photographs of completed work

5.2 BIENNIAL INSPECTION REPORT

A written report should be prepared for each biennial bridge inspection performed by certified bridge safety inspection personnel. The report requirements are described in the applicable specifications adopted by each state department of transportation (or equivalent entity) and various other transportation agencies (Turnpike Commissions, Thruway Authorities, Port Authorities, etc.). In addition to the required information delineated in the appropriate specification, the report should also include the following information about the cables and suspension system:

- Separate listings of the ratings applied to each component
- Photographs of deficiencies
- Reasons for ratings lower than 5
- Recommended action items
- Reasons for recommending an internal inspection (if applicable)

There are two major rating guideline systems currently used throughout the country (FHWA's Bridge Inspector's Reference Manual):

- FHWA's *Recording and Coding Guide for the Structural Inventory and Appraisal of the Nation's Bridges* [2] used for the National Bridge Inventory (NBI) component rating method.
    - A single-digit code for Item 59 on the Federal Structure Inventory and Appraisal (SI&A) sheet indicates the overall condition of the superstructure.
    - Condition Rating from 0 to 9, where 9 is the best possible rating.
    - Use both the current and previous inspection findings to determine the rating.
- The *AASHTO Guide Manual for Bridge Element Inspection* [8] used for the element level condition state assessment method.
    - The National Bridge Elements are:

Element #	**Description**
147	Steel Main Cables (suspension cables)
148	Secondary Steel Cables (suspender cables)

    - The quantity for Element #147 is the sum of the length of the main cables, and the quantity for Element #148 is the sum of the length of the secondary steel cables. For both Elements, the total quantity is stratified over four standard condition states comprised of good, fair, poor, and severe general descriptions.
    - For cable damage due to fatigue, use the "Steel Cracking/Fatigue" Smart Flag (Defect Flag), Bridge Management Element #356 to identify the predominate defect in a given condition state that is not corrosion.
    - For cable damage due to traffic impact, use the "Superstructure Traffic Impact (load capacity)" Smart Flag (Defect Flag), Bridge Management Element #362 to identify all traffic collisions with the superstructure. Application of the flag is in relation to the impact on the structure's capacity to carry load.
    - For cables with section loss, use the "Steel Section Loss" Smart Flag (Defect Flag), Bridge Management Element #363 to identify the predominate defect in a given condition state that is not corrosion. Setting this flag will identify the severity of section loss.

5.3 INTERNAL INSPECTION REPORT

A written report should be prepared for each internal cable inspection. These should be performed by qualified bridge inspection personnel led by a chief inspector who is a professional engineer with experience in cable inspections. The report should include:

- Executive Summary
    - Number of locations opened for inspection
    - General description of conditions found
    - Strength of each panel investigated
    - Safety factor of each panel investigated
    - Safety factor using the panel with the lowest strength and the maximum cable tension (typically located adjacent to the tower)

- - Recommendations for remedial actions
  - Recommendation for date of next inspection
- Table of Contents
- Summary addressing executive summary items in greater detail
- Findings from preliminary cable walk and reasons for selecting investigated panels
- Plan and Elevation of cables showing locations of panels investigated
- Description and photographs of the means of access to the cables
- Detailed descriptions of each panel opened
  - Cable cross-sections showing wedge locations
  - Distribution of corrosion stages
  - Location of broken wires
- Summary of laboratory test results
  - Cable cross-sections showing locations of sample wires
- Description of method used to calculate cable strength
  - Table of calculated strengths
- Table of cable tensions due to dead load, live load and temperature
  - Table of Cable Safety Factors
- Investigator's estimate of the accuracy of estimated cable strength
- Conclusions
  - Discussion of cable strengths, safety factors and possible errors
  - Discussion of probable causes of deterioration
- Recommendations
  - Plan for continued operation of the bridge if the Safety Factor is low
  - General plan for maintenance and repairs
  - Specific plan for time of next inspection and number of panels to be inspected
- Appendices
  - Laboratory Reports
    - Wire properties from tests, means and standard deviations of corrosion groups
    - Weight of Zinc Coating Test and Preece Test
    - Chemical testing of metal and corrosion products
    - Metallurgical examinations with photographs
  - Sample strength calculations
  - Photographs showing cable exterior (from cable walk) and cable interior (from wedging)
  - Photographs of inspection and rewrapping operations

The safety factor for an inspected panel is the cable strength in that panel divided by the cable tension in that panel, due to dead load, live load, and temperature effects. The safety factor of the cable when it is inspected in its entirety is the lowest value of all the panels. The safety factor of the cable when only a portion of it has been inspected is determined by using the minimum cable strength and the maximum cable tension.

6.0 REFERENCES

1. Mayrbaurl, R.M., and Camo, S., (2004a), *Guidelines for Inspection and Strength Evaluation of Suspension Bridge Parallel Wire Cables*, NCHRP Report 534, Transportation Research Board, Washington, D.C.

2. FHWA Office of Engineering, Bridge Division, Bridge Management Branch, (1995), *Recording and Coding Guide for the Structural Inventory and Appraisal of the Nation's Bridges*, Report No. FHWA-PD-96-001, Washington, D.C.

3. Ryan, T.W., Hartle, R.A., Mann, J.E., and Danovich, L.J., (2006), *Bridge Inspector's Reference Manual*, Publication No. FHWA NHI 03-001, U.S. Government Printing Office, Washington, D.C.

4. Hopwood, T. and Havens, J.H., (1984), *Corrosion of Cable Suspension Bridges*, Kentucky Transportation Research Program, University of Kentucky, Lexington, Kentucky.

5. Mayrbaurl, R.M., and Camo, S., (2004b), *Structural Safety Evaluation of Suspension Bridge Parallel-Wire Cables*, Final Report prepared for National Cooperative Highway Research Program/Transportation Research Board (available as a CD with NCHRP Report 534).

6. AASHTO, (2008), *The Manual for Bridge Evaluation, 1st Edition,* AASHTO, Washington, D.C.

7. AASHTO, (2010), *LRFD Bridge Design Specifications, 5th Edition,* AASHTO, Washington, D.C.

8. AASHTO Subcommittee on Bridges and Structures, (2011), *AASHTO Guide Manual for Bridge Element Inspection, First Edition*, American Association of State Highway and Transportation Officials, Washington, D.C.

7.0 APPENDIX A: STRENGTH EVALUATION EXAMPLE

This strength evaluation example demonstrates the use of the Simplified Strength Model, and is largely based upon the example shown in Appendix C.2 of *NCHRP Report 534* (Mayrbaurl and Camo 2004). The bridge and the data for this example are entirely fictional. The intent of the example demonstrated herein is to provide more detailed and step-by-step calculations for the Simplified Strength Model. This example is divided into four sections:
- Calculations that should be accomplished prior to Inspection.
- Analyzing the Inspection Data.
- Analyzing the Laboratory Testing Data
- Cable Strength Evaluation.

The suspension cable is assumed to consist of 37 strands, with 270 wires in each strand for a total of 9,990 wires in the cable. Each wire has a nominal diameter of 0.192 inches, before galvanizing. The galvanization thickness is not included in the strength evaluation as it is assumed that the galvanization has been lost or compromised in this example. Panel 77-78 is the evaluated panel for this example, and is one of six panels inspected: three panels on each cable were selected for inspection.

The strength evaluation employs the use of the Simplified Strength Model, as described in Article 5.3.3.1 of *NCHRP Report 534*. The Simplified Model should be applied to cables that have very few cracks, whereas the upper limit is 10% of the total wire population. The Brittle-Wire Model should be used if this upper limit is exceeded. Even though in some cases the strength may be underestimated by up to 20%, the Simplified Model can be used to locate the most severely deteriorated panel among those inspected. Then the more complex strength models (Brittle-Wire) can be employed to the worst-case panel to develop a more realistic strength estimate.

It must be noted that for this example, the Simplified Model is abridged even more by assuming that only the inspected panel is deteriorated and all other panels are perfect. Only broken and cracked wires in the inspected panel are considered, and broken wires beyond the particular panel investigated are not considered (the effective development length is taken to be 1 panel because it is assumed there are no broken wires outside the panel of interest that need to be redeveloped). In other words, the calculations presented in this example are for the mean strength of one panel, assuming the cable is safe beyond the cable bands. The intent of this example is to show the Simplified Model and its potential use to determine critical locations, not to provide a final estimate of the cable strength. A deterministic estimate of the cable strength, as provided in this example, is not necessarily useful for the final estimate of cable strength, given the uncertainties involved, and the importance of the bridge. However, the Simplified Method shown in this example can be used to quickly and easily identify the worst case of the inspected panels for more detailed analysis. More complex models should be used to provide a final estimate of the cable strength for a particular evaluation project.

7.1 PRIOR TO INSPECTION

It is typically the case that details, with regard to number of wires per cable, number of rings per cable, and the number of wires per ring, are determined prior to the cable inspection. This section will demonstrate the calculations required in order to determine these cable properties. The calculations presented in this section are based on the example calculations provide in *NCHRP Report 534*, Appendix C.

7.1.1 Number of Wires in the Cable

The number of wires is typically determined from the contract drawings. For this example:
> Number of Strands in the suspension cable = 37
> Number of Wires in each Strand = 270

Therefore, the total number of wires (N) in the suspension cable under consideration is:
> N = (37 strands/cable) * (270 wires/strand) = **9990 (wires/cable)**

For the purpose of analyzing data gathered in the field, it is assumed that the cable is composed of concentric rings of wires arranged around a central wire, as shown in Figure 7-1. It should be noted however, that the wires in a cable are not actually in precise rings, but it is assumed they are because it facilitates the estimation of the number of wires at a specific depth inside the wedged opening.

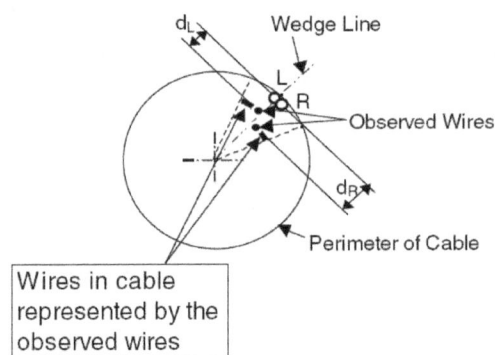

Figure 7-1 Drawing showing wires in half-sectors (taken from *NCHRP Report 534*)

7.1.2 Number of Rings in the Cable

The rings of the cable are defined as the circular portion of the cable that includes all wires at a given distance from the center of the cable. The number rings in the cable is estimated by:

$$X = \sqrt{\frac{N}{\pi}} + 0.5, \text{ rounded to the next highest integer} \qquad (4.3.1.1\text{-}1)$$

> where: X = number of rings in the cable not including the center wire
> N = actual number of wires in the cable

$$X = \sqrt{\frac{9{,}990}{\pi}} + 0.5 = 56.9$$, rounding up the nearest integer, X = 57 rings.

Therefore, there are a total of **58 rings**, including the center ring.

In most cases, the wedge lines will be equally spaced around the cable. Each sector is a pie-shaped portion of the cable, where the observed condition of the wires along the edge of the wedge represents all of the wires in the particular half-sector. For this example, as shown in Figure 7-2, the cable is divided into 8 sectors; however, this may not always be the case.

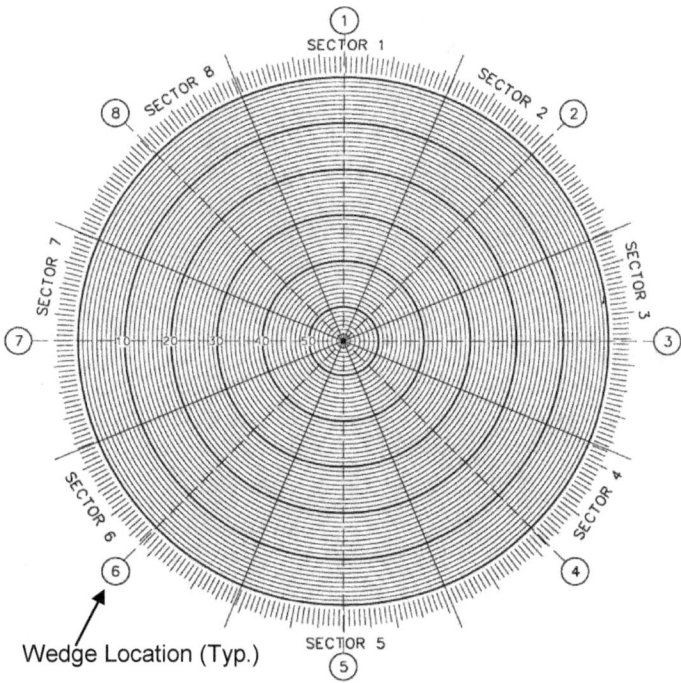

Figure 7-2 Drawing showing a cable divided into eight sectors (taken from *NCHRP Report 534*)

7.1.3 Number of Wires in Each Ring

The number of wires in each individual ring is approximated by the following fraction:

$$n_x = \frac{2x(N-1)}{X(X+1)} \qquad (4.3.1.2\text{-}1)$$

where: n_x = number of wires in ring x
x = number of rings from the center of the cable to a specific ring

The center wire in the cable lies at $x = 0$, and subsequently n_0 is taken as 1. The number of rings is not rounded, as the summation of all the rings will produce the total number of wires in the cable, N. This calculation is carried out for each individual ring.

For ring number 1, which is located at the outer edge of the cable, a distance of 57 rings from the center ring, ($x = 57$):

$$n_{57} = \frac{2(57)(9{,}990-1)}{57(57+1)} = 344.4 \text{ wires in ring } n_{57}.$$

Likewise, for ring number 20 (counting from the outside ring to the inside), a distance of 38 rings from the center ring, ($x = 38$):

$$n_{38} = \frac{2(38)(9{,}990-1)}{57(57+1)} = 229.6 \text{ wires in ring } n_{38}.$$

The above calculations are performed for each ring, and summarized in Table 7-1.

Table 7-1 Number of wires in each ring

d	x	n	d	x	n
1	57	344.4	30	28	169.2
2	56	338.4	31	27	163.2
3	55	332.4	32	26	157.1
4	54	326.3	33	25	151.1
5	53	320.3	34	24	145.0
6	52	314.2	35	23	139.0
7	51	308.2	36	22	132.9
8	50	302.1	37	21	126.9
9	49	296.1	38	20	120.9
10	48	290.1	39	19	114.8
11	47	284.0	40	18	108.8
12	46	278.0	41	17	102.7
13	45	271.9	42	16	96.7
14	44	265.9	43	15	90.6
15	43	259.8	44	14	84.6
16	42	253.8	45	13	78.6
17	41	247.8	46	12	72.5
18	40	241.7	47	11	66.5
19	39	235.7	48	10	60.4
20	38	229.6	49	9	54.4
21	37	223.6	50	8	48.3
22	36	217.5	51	7	42.3
23	35	211.5	52	6	36.3
24	34	205.5	53	5	30.2
25	33	199.4	54	4	24.2
26	32	193.4	55	3	18.1
27	31	187.3	56	2	12.1
28	30	181.3	57	1	6.0
29	29	175.2	58	0	1.0
			Total number of wires =		9,990

Where: d = Ring number from the outside of the cable.
x = Distance from the center of the cable.
n = Number of wires in the given ring.

7.2 ANALYZING INSPECTION DATA

This section provides an example as to how the observations in the field are translated into useful information for the strength evaluation of the cable. As discussed previously, the condition of the wires at the edge of the wedge section for a particular location inspected should be recorded. For example, Figure 7-3 shows the data recorded during the field inspection for the wedge opening of Sector 5. Other sectors are recorded similarly; the field inspection data are not shown in this calculation. Table 7-2 summarizes the findings for each wedged opening.

In Table 7-2, there is one line for each cable ring, and the estimated number of wires in that ring. The two columns for each sector number represent each wedge, one for the left-hand side (L) and one for the right-hand side (R). The stage of corrosion for each ring in the sector is then recorded, based on figures similar to Figure 7-3 for the other wedges.

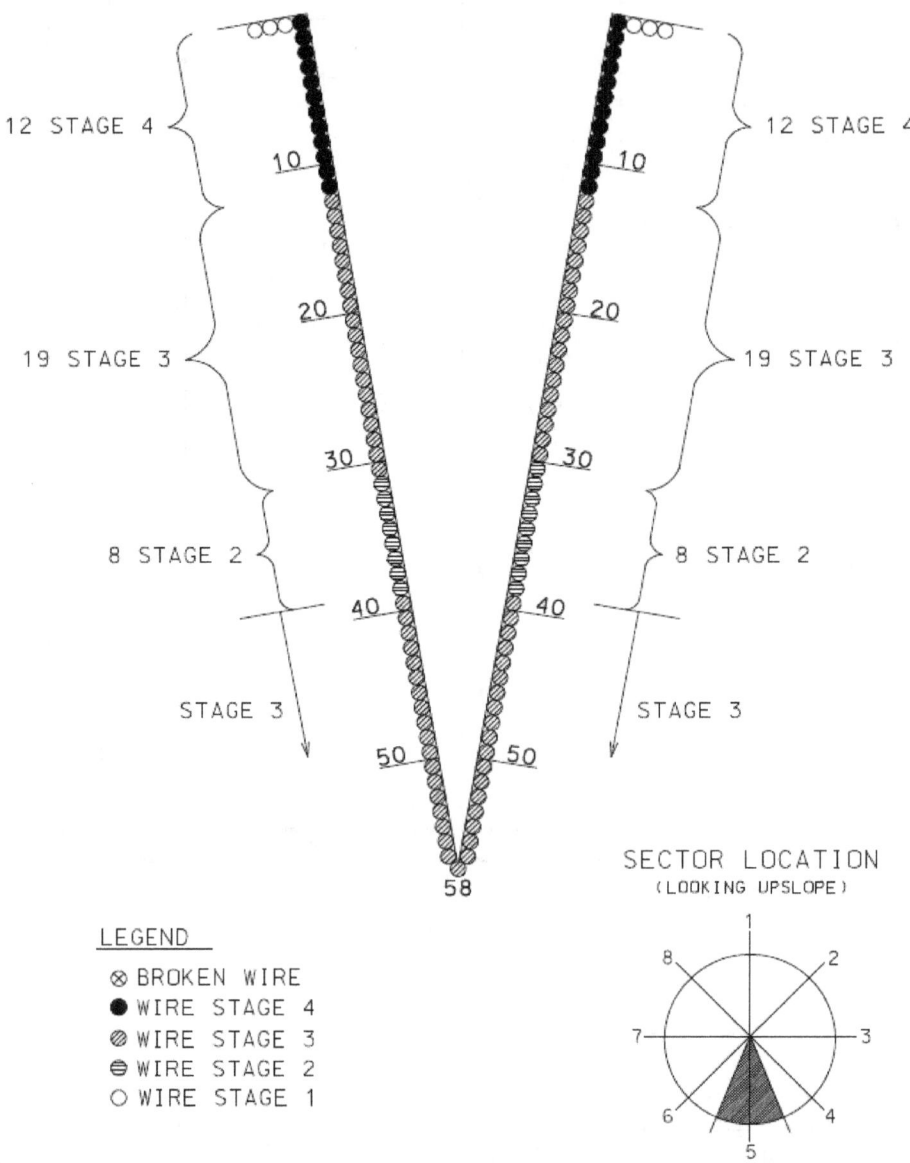

Figure 7-3 Drawing showing field inspection data for the sector five wedged opening

Table 7-2 Corrosion stages assigned to individual wires

GRADING OF DETERIORATED WIRES
(Corrosion Stages Assigned to Individual Wires)

Ring Number, d	Number of Wires in Ring, n	SECTOR NUMBER														
		1		2		3		4		5		6		7		8	
		L	R	L	R	L	R	L	R	L	R	L	R	L	R	L	R
1	344.4	4	4	4	4	4	4	4	4	4	4	4	4	4	4	4	4
2	338.4	4	4	4	4	4	4	4	4	4	4	4	4	4	4	4	4
3	332.4	4	4	4	4	4	4	4	4	4	4	4	4	4	4	4	4
4	326.3	4	4	4	4	4	4	4	4	4	4	4	4	3	3	3	3
5	320.3	3	3	4	4	4	4	4	4	4	4	4	4	3	3	3	3
6	314.2	3	3	4	4	3	3	4	4	4	4	4	4	3	3	3	3
7	308.2	3	3	4	4	3	3	4	4	4	4	4	4	3	3	3	3
8	302.1	3	3	4	4	3	3	3	3	4	4	3	3	3	3	3	3
9	296.1	3	3	3	3	3	3	3	3	4	4	3	3	3	3	3	3
10	290.1	3	3	3	3	3	3	3	3	4	4	3	3	3	3	3	3
11	284.0	3	3	3	3	3	3	3	3	4	4	3	3	3	3	3	3
12	278.0	3	3	3	3	3	3	3	3	4	4	3	3	3	3	2	2
13	271.9	3	3	3	3	3	3	3	3	4	4	3	3	3	3	2	2
14	265.9	3	3	3	3	3	3	3	3	3	3	3	3	4	4	2	2
15	259.8	3	3	3	3	3	3	4	4	3	3	4	4	2	2	2	2
16	253.8	3	3	3	3	3	3	4	4	3	3	3	3	2	2	2	2
17	247.8	3	3	3	3	3	3	3	3	3	3	3	3	2	2	2	2
18	241.7	3	3	3	3	3	3	3	3	3	3	3	3	2	2	2	2
19	235.7	3	3	3	3	3	3	3	3	3	3	3	3	2	2	2	2
20	229.6	3	3	3	3	3	3	3	3	3	3	3	3	2	2	2	2
21	223.6	3	3	3	3	3	3	3	3	3	3	3	3	2	2	2	2
22	217.5	3	3	3	3	3	3	3	3	3	3	3	3	2	2	2	2
23	211.5	3	3	3	3	3	3	3	3	3	3	3	3	2	2	2	2
24	205.5	3	3	3	3	3	3	3	3	3	3	3	3	2	2	2	2
25	199.4	2	2	3	3	3	3	3	3	3	3	3	3	2	2	2	2
26	193.4	2	2	3	3	3	3	3	3	3	3	3	3	2	2	2	2
27	187.3	3	3	3	3	3	3	3	3	3	3	2	2	3	3	3	3
28	181.3	3	3	3	3	3	3	3	3	3	3	2	2	3	3	3	3
29	175.2	3	3	3	3	3	3	3	3	3	3	2	2	2	2	2	2
30	169.2	4	4	3	3	3	3	3	3	3	3	2	2	3	3	3	3
31	163.2	2	2	3	3	3	3	3	3	3	3	2	2	2	2	2	2
32	157.1	2	2	3	3	3	3	3	3	3	3	2	2	3	3	3	3
33	151.1	2	2	2	2	3	3	3	3	2	2	2	2	3	3	2	2
34	145.0	2	2	2	2	3	3	3	3	2	2	2	2	3	3	2	2
35	139.0	2	2	2	2	3	3	3	3	2	2	2	2	3	3	2	2
36	132.9	2	2	2	2	3	3	3	3	2	2	2	2	3	3	2	2
37	126.9	2	2	2	2	3	3	3	3	2	2	2	2	3	3	2	2
38	120.9	3	3	3	3	3	3	3	3	2	2	2	2	3	3	2	2
39	114.8	2	2	2	2	3	3	3	3	2	2	2	2	3	3	2	2
40	108.8	2	2	2	2	3	3	3	3	3	3	2	2	3	3	2	2
41	102.7	2	2	2	2	2	2	3	3	3	3	2	2	2	2	2	2
42	96.7	2	2	2	2	2	2	3	3	3	3	2	2	2	2	2	2
43	90.6	2	2	2	2	2	2	3	3	3	3	2	2	3	3	3	3
44	84.6	3	3	3	3	3	3	3	3	3	3	2	2	2	2	2	2
45	78.6	3	3	3	3	3	3	3	3	3	3	2	2	2	2	2	2
46	72.5	3	3	3	3	3	3	3	3	3	3	2	2	2	2	2	2
47	66.5	2	2	2	2	2	2	3	3	3	3	2	2	2	2	2	2
48	60.4	2	2	2	2	2	2	3	3	3	3	2	2	2	2	2	2
49	54.4	2	2	2	2	2	2	3	3	3	3	2	2	2	2	2	2
50	48.3	2	2	2	2	2	2	3	3	3	3	2	2	2	2	2	2
51	42.3	2	2	2	2	2	2	3	3	3	3	2	2	2	2	2	2
52	36.3	2	2	2	2	2	2	3	3	3	3	2	2	2	2	2	2
53	30.2	2	2	2	2	2	2	3	3	3	3	2	2	2	2	2	3
54	24.2	2	2	2	2	2	2	3	3	3	3	2	2	2	2	2	2
55	18.1	2	2	2	2	2	2	3	3	3	3	2	2	2	2	2	2
56	12.1	2	2	2	2	2	2	3	3	3	3	2	2	2	2	2	2
57	6.0	2	2	2	2	2	2	3	3	3	3	2	2	2	2	2	2
58	1.0	2	2	2	2	2	2	3	3	3	3	2	2	2	2	2	2

7.2.1 Corrosion Map

The data collected in the field, as summarized in Table 7-2, can be used to develop a corrosion map of the entire cable for the selected panel point, as shown in Figure 7-4. The corrosion map provides a visual representation of the amount of corrosion in the cable, allowing the engineer to better understand what was observed in the field during the actual inspection.

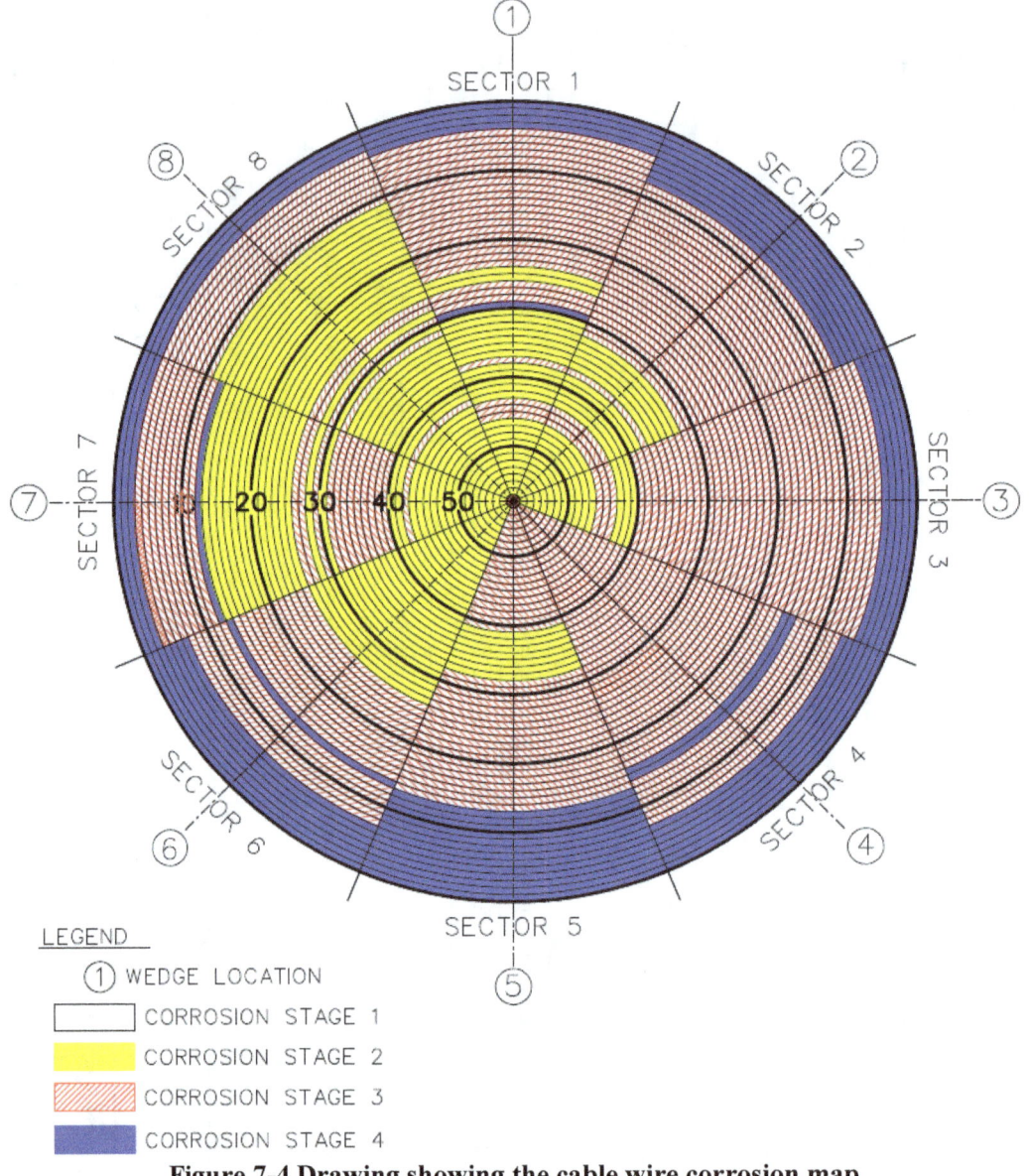

Figure 7-4 Drawing showing the cable wire corrosion map

7.2.2 Broken and Removed Wires for Testing

Another item that is recorded as part of the field inspection is the number and location of broken wires and those removed for testing. Broken wires can usually be detected in the layer of wires below the outermost layer, and may be discovered in the wedged openings as well. For this example, the number and location of the broken wires are recorded as shown in Figure 7-5. A total of 8 wires were found to be broken at this particular panel location.

The number and location of wires removed for sampling, and the stage of corrosion, should also be recorded, as shown in Figure 7-5. Furthermore, whenever a sample wire is removed from the cable for testing, the gap that forms after the first cut should be measured. The capacity of the cable band to develop wire tension can be estimated on a statistical basis from the measured gaps and the dead and assumed live load at the time of measurement.

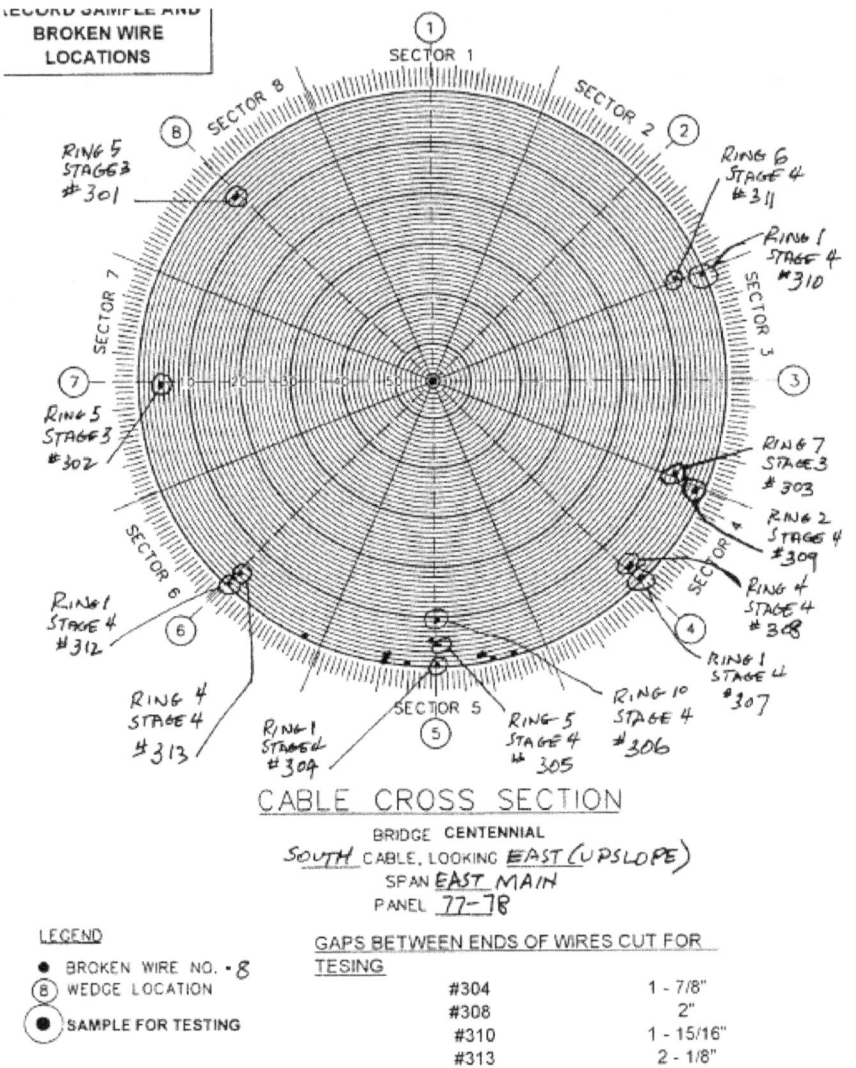

Figure 7-5 Drawing showing the map of broken wires and wires removed for testing (taken from *NCHRP Report 534*)

7.2.3 Number of Wires in Each Corrosion Stage

For each ring in the cable, the number of wires in each particular stage of corrosion can be estimated based on the findings for each wedged opening and half-sector, as summarized in Table 7-2. In a given ring, the number of wires classified under each particular corrosion stage is determined based on the percentage of corrosion stages found in the wedged openings.

For example, as shown in Table 7-2, the following data for Ring Number 14 has been recorded:
- total number of wires in the ring is 265.9 ($n_{14} = 265.9$)
- 0 of the 16 half-sectors are classified as Stage 1
- 2 of the 16 half-sectors are classified as Stage 2
- 12 of the 16 half-sectors are classified as Stage 3
- 2 of the 16 half-sectors are classified as Stage 4.

Using the above data, the number of wires in each corrosion stage in Ring Number 14 can be determined from simple ratios, as follows:

Number of wires in Stage 1:

$$N_{s1} = \frac{\text{Number of Half - Sectors with Stage 1 Corrosion}}{\text{Total Number of Half - Sectors}} \cdot (\text{Total Number of Wires in Ring 14})$$

$$N_{s1} = \frac{0}{16} \cdot (265.9) = 0.0 \text{ Wires in Stage 1}$$

Number of wires in Stage 2:

$$N_{s2} = \frac{\text{Number of Half - Sectors with Stage 2 Corrosion}}{\text{Total Number of Half - Sectors}} \cdot (\text{Total Number of Wires in Ring 14})$$

$$N_{s2} = \frac{2}{16} \cdot (265.9) = 33.2 \text{ Wires in Stage 2}$$

Number of wires in Stage 3:

$$N_{s3} = \frac{\text{Number of Half - Sectors with Stage 3 Corrosion}}{\text{Total Number of Half - Sectors}} \cdot (\text{Total Number of Wires in Ring 14})$$

$$N_{s3} = \frac{12}{16} \cdot (265.9) = 199.4 \text{ Wires in Stage 3}$$

Number of wires in Stage 4:

$$N_{s4} = \frac{\text{Number of Half - Sectors with Stage 4 Corrosion}}{\text{Total Number of Half - Sectors}} \cdot (\text{Total Number of Wires in Ring 14})$$

$$N_{s4} = \frac{2}{16} \cdot (265.9) = 33.2 \text{ Wires in Stage 4}$$

The above calculations employ the fact that 8 wedged openings (16 half-sectors) were recorded during the inspection. If fewer sectors are utilized, this must be taken into account in these calculations. Furthermore, the above calculations assume that the all sectors are of relatively equal size (as best as physically possible in the field). However, if the sectors are not the same size, the above calculations need to be modified accordingly.

Calculations similar to those shown for Ring Number 14 are performed for all the other rings, and are summarized in Table 7-3.

Furthermore, the total number of wires in each corrosion stage for the entire cross section of cable is calculated in Table 7-3. There are zero wires in corrosion Stage 1; 2273.7 wires in corrosion Stage 2; 5540.9 wires in corrosion Stage 3; and 2175.5 wires in corrosion Stage 4. These values result in the following percentages of wires in each corrosion stage: 0.0%, 22.8%, 55.5%, and 21.8% for corrosion Stages 1, 2, 3, and 4, respectively.

Table 7-3 Number of wires in each corrosion stage

Ring Number, d	Number of Wires in Ring, n	Number of Half-Sectors in Each Corrosion Stage Stage Number, k				Number of Wires in Each Corrosion Stage, N_{sk} Stage Number, k			
		1	2	3	4	1	2	3	4
1	344.4	0	0	0	16	0.0	0.0	0.0	344.4
2	338.4	0	0	0	16	0.0	0.0	0.0	338.4
3	332.4	0	0	0	16	0.0	0.0	0.0	332.4
4	326.3	0	0	4	12	0.0	0.0	81.6	244.7
5	320.3	0	0	6	10	0.0	0.0	120.1	200.2
6	314.2	0	0	8	8	0.0	0.0	157.1	157.1
7	308.2	0	0	8	8	0.0	0.0	154.1	154.1
8	302.1	0	0	12	4	0.0	0.0	226.6	75.5
9	296.1	0	0	14	2	0.0	0.0	259.1	37.0
10	290.1	0	0	14	2	0.0	0.0	253.8	36.3
11	284.0	0	0	14	2	0.0	0.0	248.5	35.5
12	278.0	0	2	12	2	0.0	34.7	208.5	34.7
13	271.9	0	2	12	2	0.0	34.0	203.9	34.0
14	**265.9**	**0**	**2**	**12**	**2**	**0.0**	**33.2**	**199.4**	**33.2**
15	259.8	0	4	8	4	0.0	65.0	129.9	65.0
16	253.8	0	4	10	2	0.0	63.5	158.6	31.7
17	247.8	0	4	12	0	0.0	61.9	185.8	0.0
18	241.7	0	4	12	0	0.0	60.4	181.3	0.0
19	235.7	0	4	12	0	0.0	58.9	176.8	0.0
20	229.6	0	4	12	0	0.0	57.4	172.2	0.0
21	223.6	0	4	12	0	0.0	55.9	167.7	0.0
22	217.5	0	4	12	0	0.0	54.4	163.2	0.0
23	211.5	0	4	12	0	0.0	52.9	158.6	0.0
24	205.5	0	4	12	0	0.0	51.4	154.1	0.0
25	199.4	0	6	10	0	0.0	74.8	124.6	0.0
26	193.4	0	6	10	0	0.0	72.5	120.9	0.0
27	187.3	0	2	14	0	0.0	23.4	163.9	0.0
28	181.3	0	2	14	0	0.0	22.7	158.6	0.0
29	175.2	0	6	10	0	0.0	65.7	109.5	0.0
30	169.2	0	2	12	2	0.0	21.2	126.9	21.2
31	163.2	0	8	8	0	0.0	81.6	81.6	0.0
32	157.1	0	4	12	0	0.0	39.3	117.8	0.0
33	151.1	0	10	6	0	0.0	94.4	56.7	0.0
34	145.0	0	10	6	0	0.0	90.6	54.4	0.0
35	139.0	0	10	6	0	0.0	86.9	52.1	0.0
36	132.9	0	10	6	0	0.0	83.1	49.9	0.0
37	126.9	0	10	6	0	0.0	79.3	47.6	0.0
38	120.9	0	6	10	0	0.0	45.3	75.5	0.0
39	114.8	0	10	6	0	0.0	71.8	43.1	0.0
40	108.8	0	8	8	0	0.0	54.4	54.4	0.0
41	102.7	0	12	4	0	0.0	77.0	25.7	0.0
42	96.7	0	12	4	0	0.0	72.5	24.2	0.0
43	90.6	0	8	8	0	0.0	45.3	45.3	0.0
44	84.6	0	6	10	0	0.0	31.7	52.9	0.0
45	78.6	0	6	10	0	0.0	29.5	49.1	0.0
46	72.5	0	6	10	0	0.0	27.2	45.3	0.0
47	66.5	0	12	4	0	0.0	49.9	16.6	0.0
48	60.4	0	12	4	0	0.0	45.3	15.1	0.0
49	54.4	0	12	4	0	0.0	40.8	13.6	0.0
50	48.3	0	12	4	0	0.0	36.3	12.1	0.0
51	42.3	0	12	4	0	0.0	31.7	10.6	0.0
52	36.3	0	12	4	0	0.0	27.2	9.1	0.0
53	30.2	0	12	4	0	0.0	22.7	7.6	0.0
54	24.2	0	12	4	0	0.0	18.1	6.0	0.0
55	18.1	0	12	4	0	0.0	13.6	4.5	0.0
56	12.1	0	12	4	0	0.0	9.1	3.0	0.0
57	6.0	0	12	4	0	0.0	4.5	1.5	0.0
58	1.0	0	12	4	0	0.0	0.8	0.3	0.0
$\Sigma n = N =$	9990.0			TOTAL = N_{sk} =		0.0	2273.7	5540.9	2175.5
				$N_{sk}/N =$		0.0%	22.8%	55.5%	21.8%

7.3 ANALYZING LABORATORY TESTING DATA

Several sample wires are typically taken from all of the panel locations inspected. These samples, which encompass various stages of corrosion, are tested for wire strength via tensile tests, as discussed in section 2 of this Primer. The tensile test data is then used to determine the tensile strength distribution of each corrosion group of wires, and the subsequent strength of the entire cable.

For this particular example, several sample wires were removed throughout the cable for testing. Ten corrosion stage 1, 15 stage 2, 18 stage 3, and 30 stage 4 wires were selected for testing. It should be noted that although the deterioration of the cable was observed to be severe in Panel 77-78, the limited number of panels opened in the inspection was insufficient for a larger sampling of wires to be tested. Given the condition of the observed panel, additional wires should be sampled from the next inspection.

7.3.1 Test data for Wire Number 609

Tension tests results of single wire sample number 609 are shown in Table 7-4. It is noted that the test results shown in Table 7-4 are for a wire taken from panel 76-77, of the North cable, and not from the panel used throughout this example: Panel 77-78 of the South cable. The following calculations would be the same regardless of the wire sample selected.

Wire sample number 609 was divided into 11 segments. Each specimen has a length of 12 inches, as measured between the grips of the tensile test machine. As shown in Table 7-4, the mean tensile strength (μ_{sj}) is calculated as 239.8 ksi, with a standard deviation (σ_{sj}) of 0.5. From this data, the probable minimum tensile strength of the wire can be calculated, in accordance with equation 4.4.3.2-1 of *NCHRP Report 534*:

$$x_{1,j} = \mu_{sj} + \Phi^{-1}\left(\frac{L_0}{L}\right) \cdot \sigma_{sj} \qquad (4.4.3.2\text{-}1)$$

where: $x_{1,j}$ = probable minimum value of x_j in a length L of the wire from which sample j is removed
$\Phi^{-1}(L_0/L)$ = inverse of the standard normal cumulative distribution for the probability L_0/L, determined using Figure 7-6 (Figure 4.4.3.2-1 in *NCHRP Report 534*).
L_0 = length of test specimen between the grips of the testing machine
L = length of wire between centers of cable bands

Given that length of the test specimens between the grips of the testing machine is 12 in. (L_0), and the length of panel in the actual structure is 41 ft (L),

$$\frac{L}{L_0} = \frac{492 \text{in.}}{12 \text{in.}} = 41$$

Using the above value, and Figure 7-6, the term $\Phi^{-1}(L_0/L)$ is found to be -1.97.

Table 7-4 Results of tension tests for wire number 609

CABLE AND PANEL =	NWM7677
WIRE SAMPLE =	609

Stresses Based on:	
Diameter of wire =	0.192 in
Area of wire =	0.028953 in²

CABLE AND PANEL DESIGNATIONS
W = WEST   M = MAIN
E = EAST   S = SIDE
N = NORTH CABLE   2123 = PANEL 21-23
S = SOUTH CABLE   ANCH = ANCHORAGE

LEGEND
X = NUMBER OF SAMPLE WIRE
X.01 = SPECIMEN No. 1 FROM WIRE X
X.02 = SPECIMEN No. 2 FROM WIRE X
X.91 = LONG SPECIMEN FROM WIRE X

Specimen Number	Corrosion Stage	Max Load (lbs)	Yield Strength (ksi)	Tensile Strength (ksi)	Elongation in 10" (%)	Reduction in Area (%)	Remarks	Fracture Type
609.01	3	6918	200.0	238.9	5	29.00	Note 2, L	B
609.02	3	6938	201.0	239.6	4.5	35.50	Note 2, L	B
609.03	3	6934	200.0	239.5	5	27.00	Note 2, H	B
609.04	3	6938	200.0	239.6	5.5	35.50	Note 2, M	B
609.05	3	6930	200.0	239.4	6	39.00	Note 2, M	B
609.06	3	6930	200.0	239.4	5	29.50	Note 2, L	B
609.07	3	6950	200.0	240.0	5	37.50	Note 2, L	B
609.08	3	6962	201.0	240.5	5	37.50	Note 2, L	B
609.09	3	6962	201.0	240.5	5.5	39.00	Note 2, L	B
609.10	3	6958	202.0	240.3	5	37.50	Note 2, L	B
609.11	3	6954	200.0	240.2	6	31.50	Note 2, M	B
11	= number of samples							
Mean		6943		239.8	5.2	34.41		
Standard Deviation		15		0.5	0.5	4.36		
Maximum				240.5				
Minimum				238.9				

FRACTURE TYPES
A	Ductile; Cup and cone
B	Ductile; Cup and cone with shear lips alternating above and below fracture plane
B-C	Semi-ductile; Ragged with partial shear lips and reduced reduction in area
C	Brittle; Ragged with minimal or no reduction in area
D	Brittle w/Crack; Fracture with partial crack

NOTE 1: Surface of wire covered with a gummy material, possibly dried linseed oil
NOTE 2: Surface corrosion is present at the fracture location, which is the probable initiation point of the fracture
   L = LOCAL   M = MODERATE   S = SEVERE
   O = OVERALL   H = HEAVY

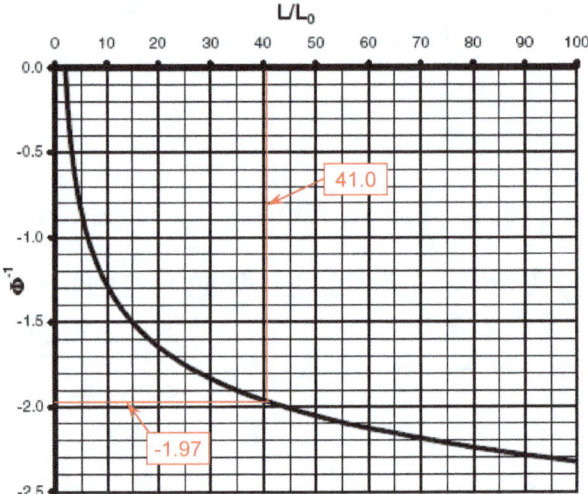

Figure 7-6 Graph for computing the inverse of the standard normal cumulative distribution

The above known values are then substituted into equation 4.4.3.2-1 to determine the probable minimum strength of the particular wire sample, in this case wire sample number 609:

$$x_{1,609} = \mu_{s609} + \Phi^{-1}\left(\frac{L_0}{L}\right) \cdot \sigma_{s609} = 239.8 \text{ ksi} + (-1.97)(0.5 \text{ ksi}) = 238.8 \text{ ksi}$$

Therefore, the probable minimum tensile strength for wire sample 609 is 238.8 ksi. Wire 609 is one of the 75 samples that were tested. All of the tested wires will have similar data, and probable minimum tensile strengths determined.

7.3.2 Test Data for Wire Number 613

Tension tests results of a single wire sample number 613 are shown in Table 7-5. As noted in Table 7-5, specimens 613.01 and 613.04 both failed due to fracture associated with a partial crack in the wire. When a crack is present in one or more of the specimens of a particular wire, the calculation of the probable minimum tensile strength in accordance with equation 4.4.3.2-1 is not valid. Instead, the lowest strength determined for a cracked wire specimen is to be used. Therefore, based on wire specimen 613.01, the probable minimum tensile strength for wire 613 is taken as 208.8 ksi ($x_{1,613} = 208.8 \text{ ksi}$).

Table 7-5 Results of tension tests for wire number 613

CABLE AND PANEL =	NWM7677
WIRE SAMPLE =	613

Stresses Based on:	
Diameter of wire =	0.192 in
Area of wire =	0.028953 in^2

CABLE AND PANEL DESIGNATIONS
W = WEST   M = MAIN
E = EAST   S = SIDE
N = NORTH CABLE   2123 = PANEL 21-23
S = SOUTH CABLE   ANCH = ANCHORAGE

LEGEND
X = NUMBER OF SAMPLE WIRE
X.01 = SPECIMEN No. 1 FROM WIRE X
X.02 = SPECIMEN No. 2 FROM WIRE X
X.91 = LONG SPECIMEN FROM WIRE X

Specimen Number	Corrosion Stage	Max Load (lbs)	Yield Strength (ksi)	Tensile Strength (ksi)	Elongation in 10" (%)	Reduction in Area (%)	Remarks	Fracture Type
613.01	4	6044	198.0	208.8	0.5	2.00	Crack 1/6D, Note 2, L	D
613.02	4	6974	199.0	240.9	4.0	24.50	Note 2, M	B
613.03	4	7003	199.0	241.9	4.0	26.00	Note 2, H	B
613.04	4	6113	202.0	211.1	0.5	2.00	Crack 1/8D, Note 2, M	D
613.05	4	7017	201.0	242.4	4.5	20.00	Note 2, H	B-C
613.06	4	7005	203.0	241.9	4.0	20.00	Note 2, H	B
613.07	4	6906	200.0	238.5	5.0	21.50	Note 2, H	B-C
613.08	4	7051	202.0	243.5	3.5	24.50	Note 2, M	B-C
613.09	4	6970	201.0	240.7	3.0	20.00	Note 2, H	B-C
613.10	4	6998	201.0	241.7	4.5	29.50	Note 2, M	B
613.11	4	7012	203.0	242.2	3.5	20.00	Note 2, H	B-C

11 = number of samples

Mean		6827		235.8	3.4	19.09
Standard Deviation		372		12.8	1.5	8.99
Maximum				243.5		
Minimum				208.8		

FRACTURE TYPES
A   Ductile; Cup and cone
B   Ductile; Cup and cone with shear lips alternating above and below fracture plane
B-C Semi-ductile; Ragged with partial shear lips and reduced reduction in area
C   Brittle; Ragged with minimal or no reduction in area
D   Brittle w/Crack; Fracture with partial crack

NOTE 1: Surface of wire covered with a gummy material, possibly dried linseed oil
NOTE 2: Surface corrosion is present at the fracture location, which is the probable initiation point of the fracture
           L = LOCAL   M = MODERATE   S = SEVERE
           O = OVERALL   H = HEAVY

7.3.3 Summary of all Test Data

Probable minimum tensile strengths are determined for all wires tested, with calculations similar to those shown in Table 7-4 and Table 7-5. A summary of all the wire samples tested is provided in Table 7-6, where the samples are organized by wire property groups in accordance with Article 4.4.1 of *NCHRP Report 534*. These groups are as follows:

Group 1 - the sample exhibits Stage 1 corrosion
Group 2 - the sample exhibits Stage 1 or Stage 2 corrosion
Group 3 - the sample exhibits Stage 3 corrosion with no cracks in the wire
Group 4 - the sample exhibits Stage 4 corrosion with no cracks in the wire
Group 5 - the sample exhibits Stage 3 or Stage 4 corrosion with cracks in the wire.

Table 7-6 Summary of all tension test results

CABLE AND PANEL = BOTH
WIRE SAMPLE = ALL
YEAR = 2000

Stresses Based on:
Diameter of wire = 0.192 in
Area of wire = 0.028953 in^2

CABLE AND PANEL DESIGNATIONS
W = WEST   M = MAIN
E = EAST   S = SIDE
N = NORTH CABLE   2123 = PANEL 21-23
S = SOUTH CABLE   ANCH = ANCHORAGE

Cable and Panel	Wire Sample Number	Corr. Stage	Wire Group	Max Load (lbs)	Tensile Strength (ksi)	Remarks	Fracture Type	STRENGTH BY GROUP 1	2	3	4	5
SES0001	101	1	1	6885	237.8		B	237.8				
SES0001	102	1	1	6920	239.0		A	239.0				
SWS1718	201	1	1	6450	222.8		B	222.8				
SWS1718	202	1	1	6821	235.6		B	235.6				
NWS0001	401	1	1	6955	240.2		B	240.2				
NWS0001	402	1	1	6833	236.0		A	236.0				
NWM5758	501	1	1	6865	237.1		B	237.1				
NWM5758	502	1	1	6891	238.0		A	238.0				
NWM7677	601	1	1	6914	238.8		B	238.8				
NWM7677	602	1	1	6926	239.2		A	239.2				
SES0001	103	2	2	6865	237.1		B		237.1			
SES0001	104	2	2	6876	237.5		A		237.5			
SES0001	105	2	2	6824	235.7		A		235.7			
SWS1718	203	2	2	6969	240.7		A		240.7			
SWS1718	204	2	2	6983	241.2		A		241.2			
SWS1718	205	2	2	7091	244.9		A		244.9			
NWS0001	403	2	2	6954	240.2		A		240.2			
NWS0001	404	2	2	6946	239.9		B		239.9			
NWS0001	405	2	2	6914	238.8		B-C		238.8			
NWM5758	503	2	2	6940	239.7		B		239.7			
NWM5758	504	2	2	7004	241.9		B		241.9			
NWM5758	505	2	2	7001	241.8		B		241.8			
NWM7677	603	2	2	7001	241.8		B		241.8			
NWM7677	604	2	2	7004	241.9		B		241.9			
NWM7677	605	2	2	7148	246.9		A		246.9			
SES0001	106	3	3	6471	223.5		B-C			223.5		
SES0001	107	3	3	7029	242.8					242.8		
SES0001	108	3	3	6923	239.1		B			239.1		
SWS1718	206	3	3	6740	232.8		B			232.8		
SWS1718	207	3	3	6847	236.5		B			236.5		
SWS1718	208	3	3	6998	241.7		B			241.7		
SEM7778	301	3	3	6891	238.0		A			238.0		
SEM7778	302	3	3	6700	231.4		B			231.4		
NWM5758	506	3	3	6752	233.2		B-C			233.2		
NWM5758	507	3	3	6630	229.0		B			229.0		
NWM5758	508	3	3	6986	241.3		B			241.3		
NWM5758	510	3	3	6883	237.7		B			237.7		
NWM7677	606	3	3	6662	230.1		B			230.1		
NWM7677	608	3	3	7027	242.7		B			242.7		
NWM7677	**609**	**3**	**3**	**6914**	**238.8**		**B**			**238.8**		
NWM7677	617	3	3	6991	241.5		B			241.5		
NWM7677	618	3	3	6659	230.0		B			230.0		

Table 7-6 (continued) Summary of all tension test results

Cable and Panel	Wire Sample Number	Corr. Stage	Wire Group	Max Load (lbs)	Tensile Strength (ksi)	Remarks	Fracture Type	STRENGTH BY GROUP				
								1	2	3	4	5
SES0001	109	4	4	6346	219.2		B-C				219.2	
SES0001	111	4	4	6907	238.6						238.6	
SES0001	112	4	4	6764	233.6		B				233.6	
SEM7778	305	4	4	6775	234.0		B				234.0	
SEM7778	307	4	4	6754	233.3		B				233.3	
SEM7778	309	4	4	6868	237.2		B				237.2	
SEM7778	310	4	4	6280	216.9		C				216.9	
NWM5758	511	4	4	6483	223.9		C				223.9	
NWM5758	514	4	4	6859	236.9		B				236.9	
NWM5758	516	4	4	6920	239.0		B-C				239.0	
NWM5758	518	4	4	6931	239.4		B				239.4	
NWM7677	610	4	4	6862	237.0		B-C				237.0	
NWM7677	611	4	4	6920	239.0		B-C				239.0	
NWM7677	614	4	4	6370	220.0		C				220.0	
NWM7677	615	4	4	6320	218.3		A				218.3	
SES0001	110	4	5	6749	233.1	CRACK 0 05D	C					233.1
SES0001	113	4	5	6502	224.6	CRACK <0.1D	D					224.6
SEM7778	304	4	5	5620	194.1	CRACK 0 2D	D					194.1
SEM7778	306	4	5	5450	188.2	CRACK 0 25D	D					188.2
SEM7778	308	4	5	6211	214.5	CRACK 0.1D	D					214.5
SEM7778	311	4	5	5539	191.3	CRACK 0 2D	D					191.3
SEM7778	312	4	5	4100	141.6	CRACK 0 25D	D					141.6
SEM7778	313	4	5	6610	228.3	CRACK 0 08D	B-C					228.3
NWM5758	512	4	5	6509	224.8	CRACK 0.1D	B					224.8
NWM5758	513	4	5	4551	157.2	CRACK 0 2D	D					157.2
NWM5758	515	4	5	5220	180.3	CRACK 0.15D	D					180.3
NWM5758	517	4	5	5671	195.9	CRACK 1/8D	B-C					195.9
NWM7677	612	4	5	6323	218.4	CRACK 0 05D	B					218.4
NWM7677	**613**	**4**	**5**	**6045**	**208.8**	**CRACK 1/6D**	**C**					**208.8**
NWM7677	616	4	5	5981	206.6	CRACK 0 05D	B-C					206.6
SEM7778	303	3	NOT USED	6671	230.4	CRACK 0.15D	B					
NWM5758	509	3	NOT USED	6480	223.8	CRACK 0 2D	D					
NWM7677	607	3	NOT USED	5892	203.5	CRACK 0.15D	D					
75	= number of samples				Mean			236.5	240.7	235.9	231.1	200.5
					Standard Deviation			5.0	2.9	5.7	8.7	26.3
					GROUP Mean			239.0		235.9	231.1	200.5
					GROUP Standard Deviation			4.3		5.7	8.7	26.3

FRACTURE TYPES	
A	Ductile; Cup and cone
B	Ductile; Cup and cone with shear lips alternating above and below fracture plane
B-C	Semi-ductile; Ragged with partial shear lips and reduced reduction in area
C	Brittle; Ragged with minimal or no reduction in area
D	Brittle w/Crack; Fracture with partial crack

As shown in Table 7-6, the tensile strength mean and standard deviation of each group is determined. Therefore, for wires classified as Group 1 or 2 wires, their probable minimum tensile strength is taken as 239.0 ksi; likewise, for Group 3 wires, the minimum tensile strength of 235.9 ksi; Group 4 is 231.1 ksi; and Group 5 is 200.5 ksi.

7.3.4 Cracked Wires as a Separate Group

The fraction of cracked wires for each stage of corrosion also needs to be determined, as it will be used in later strength calculations. The fraction of cracked wires ($p_{c,k}$) is calculated in accordance with Article 4.4.2 of *NCHRP Report 534*, and is given by:

$$p_{c,k} = \frac{\text{number of cracked Stage k sample wires}}{\text{total number of Stage k sample wires}} \quad (4.4.2\text{-}1)$$

and for Stage 3 wire in particular:

$$p_{c,3} = \frac{0.33 \cdot \text{number of cracked Stage k sample wires}}{\text{total number of Stage k sample wires}} \quad (4.4.2\text{-}2)$$

If cracks are found in corrosion Stage 3 samples, these wires are usually located in the outermost layers of the cable, with corrosion Stage 4 wires nearby. The 0.33 factor in equation 4.4.2-2 adjusts for the fact that Stage 3 wires found deeper in the cable rarely exhibit cracks. If these wires do exhibit cracks, the 0.33 factor should be increased accordingly.

Therefore, based on testing results summarized in Table 7-6, the fraction of Stage 3 and Stage 4 cracked wires is as calculated as follows:

Stage 3:

$$p_{c,3} = \frac{0.33 \cdot \text{number of cracked Stage k sample wires}}{\text{total number of Stage k sample wires}} = \frac{0.33\,(3)}{20} = 0.05$$

Stage 4:

$$p_{c,4} = \frac{\text{number of cracked Stage k sample wires}}{\text{total number of Stage k sample wires}} = \frac{15}{30} = 0.50$$

7.4 CABLE STRENGTH EVALUATION – SIMPLIFIED MODEL

The Simplified Model, as provided in Article 5.3.3.1 of *NCHRP Report 534*, allows the engineer to quickly and easily identify which panels require more detailed calculations to determine the cable's strength. For this example, the Simplified Model is simplified even more by assuming that only the inspected panel is deteriorated and all other panels are perfect (the effective development length is taken as 1 panel). Only broken and cracked wires in the inspected panel are considered. More detailed calculations employing the Brittle-Wire Model or the Limited Ductility Model, as discussed in Articles 5.3.3.2 and 5.3.3.3, respectively, of *NCHRP Report 534*, can be applied to the most severely deteriorated panel to develop a more realistic estimate of the cable's strength.

7.4.1 Estimate of Number of Broken Wires in the Development Length (Panel)

In the inspected panel 77-78, of the East Main Span – South Cable, broken wires were found only in the outer layers, with none found more than six layers into the cable. Five broken wires were found in the outer ring of the cable. Also, a total of six wires were repaired. Again, adjacent panels are assumed to be perfect in this simplified technique.

The estimated number of broken wires in panel 77-78 is calculated in accordance with equation 4.3.3.2-1 of *NCHRP Report 534*. The resulting value is rounded to the higher integer, as there can not be a fraction of broken wires. The estimated number of broken wires in the panel is calculated as:

$$n_{bi} = n_{b1,i} \cdot \frac{d_0}{2} \qquad (4.3.3.2\text{-}1)$$

where: $n_{b1,i}$ = number of broken wires in the outer ring of the cable in panel i
d_0 = depth into the cable at which no broken wires were found
i = number of evaluated panels, in this case $i = 1$

and,

$$n_{b1} = n_{b1,1} \cdot \frac{d_0}{2} = (5) \cdot \frac{7}{2} = 18 \text{ wires}$$

Since only the inspected panel is being considered in this simplified technique, the effective development length is equal to one panel, and the total number of broken wires is calculated as:

$$N_b = L_e \cdot n_{b1} \qquad (5.3.2.1\text{-}2)$$

where: N_b = number of broken wires in the effective development length
n_{b1} = number of broken wires in the evaluated panel
L_e = number of panels in the effective development length, $L_e = 1$

and,

$N_b = (1) \cdot (18 \text{ wires}) = 18 \text{ wires}$

As stated previously, six wires were repaired in this panel. Since the evaluated panel is the only panel that is being considered, then the number of repaired wires is taken as:

$$N_r = n_{r1} \qquad (5.3.2.2\text{-}2)$$

where: N_r = number of broken wires that are repaired in the effective development length
n_{r1} = number of broken wires that are repaired in the evaluated panel ($i = 1$)

and,

$N_r = 6$ wires

The net number of broken wires in the effective development length is simply calculated as:

$N_b - N_r = 18$ wires $- 6$ wires $= 12$ wires

7.4.2 Cracked Wires in the Evaluated Panel

The number of cracked wires in the evaluated panel (panel 77-78) is the next item to be determined. First, the number of unbroken wires needs to be calculated in accordance with Article 5.3.2.3 of *NCHRP Report 534*. Since the net number of broken wires in the effective development length (*$N_b - N_r = 12$ wires*) is less than number of wires in corrosion stage 4 (*$N_{s4} = 2175$ wires*), the number of unbroken wires (*N_{0k}*) is calculated as:

$N_{04} = N_{s4} - N_b + N_r = 2175 - 18 + 6 = 2163$ wires (5.3.2.3-1)

$N_{03} = N_{s3} = 5541$ wires (5.3.2.3-2)

$N_{02} = N_{s2} + N_{s1} = 0 + 2274 = 2274$ wires (5.3.2.3-3)

The fraction of Stage 3 and Stage 4 cracked wires ($p_{c,3}$ and $p_{c,4}$, respectively) was calculated previously as:

$p_{c,3} = 0.05$; and $p_{c,4} = 0.50$.

Since the effective development length for this example is the evaluated panel, the number of cracked wires for corrosion Stages 3 and 4 is calculated as (rounded to the next higher integer):

Stage 3: $N_{c,3} = (p_{c,3})(N_{03}) = (0.05)(5541 \text{ wires}) = 278$ wires

Stage 4: $N_{c,4} = (p_{c,4})(N_{04}) = (0.50)(2163 \text{ wires}) = 1082$ wires

Therefore, the total number of cracked wires in the cable at this panel (panel 77-78) is taken as 1360 wires. These calculations for the number of cracked wires are summarized in Table 7-7.

Table 7-7 Cracked wires in the evaluated panel

Corrosion Stage k	Number of Wires in Each Stage N_{sk}	Net Number of Broken Wires $N_b - N_r$	Number of Unbroken Wires N_{ok}	Fraction of Unbroken Wires p_{ok}	Fraction of Cracked Wires $p_{c,k}$	Cracked Wires in Evaluated Panel $N_{c,k}$
1	0		0	0.000	0.00	0
2	2274		2274	0.228	0.00	0
3	5541		5541	0.555	0.05	278
4	2175	12	2163	0.217	0.50	1082
totals =>	9990	12	9978	1.000		1360

$N = 9990$ wires
$N_{eff} = 9978$ wires
$N_5 = \Sigma N_{c,k} = 1360$ wires

7.4.3 Estimate of Cable Strength

The fraction of the cable represented by the unbroken wires (Groups 2, 3, and 4) is combined with the wire sample mean values of minimum tensile strength derived from the testing of the representative specimens to determine the cable mean strength for the particular panel being considered. For the Simplified Model, the estimated number of cracked wires (Group 5) and broken wires are omitted from the calculation of the cable strength.

In accordance with Article 5.3.3.1.1 of *NCHRP Report 534*, the mean tensile strength and standard deviation of the cable strength are determined using the equations:

$$\mu_s = \sum_{k=2}^{4}(p_{uk} \cdot \mu_{sk}) \qquad (5.3.3.1.1\text{-}1)$$

$$\sigma_s = \sqrt{\left(\sum_{k=2}^{4} p_{uk}(\sigma_{sk}^2 + \mu_{sk}^2)\right) - \mu_s^2} \qquad (5.3.3.1.1\text{-}2)$$

in which,

$$p_{uk} = \frac{N_k}{N_{eff} - N_5} \qquad (5.3.3.1.1\text{-}3)$$

and where:
- μ_s = sample mean tensile strength of the combined groups, excluding cracked wires
- μ_{sk} = sample mean tensile strength of Group k wires (see Table 7-6)
- σ_s = sample standard deviation of the tensile strength of the combined groups of wires, excluding cracked wires
- σ_{sk} = sample standard deviation of the tensile strength of Group k wires
- p_{uk} = fraction of unbroken wires in the cable section represented by Group k
- k = corrosion stage of a group of wires ($k = 2, 3,$ and 4)

N_{eff} = effective number of unbroken wires n the evaluated panel
N_5 = number of discrete cracked wires in the effective development length
N_k = number of Group k wires in the evaluated panel

The value of p_{uk} is calculated for each wire group as follows:

$$\text{Group 2:} \quad p_{u2} = \frac{N_2}{N_{eff} - N_5} = \frac{2274}{9978 - 1360} = 0.2638$$

$$\text{Group 3:} \quad p_{u3} = \frac{N_3}{N_{eff} - N_5} = \frac{5263}{9978 - 1360} = 0.6107$$

$$\text{Group 4:} \quad p_{u4} = \frac{N_4}{N_{eff} - N_5} = \frac{1081}{9978 - 1360} = 0.1254$$

For Wire Group 2, the mean tensile strength (μ_{s2}) and the standard deviation of the tensile strength (σ_{s2}) have been previously determined to be 239.0 ksi and 4.3 ksi, respectively (see Table 7-6). Therefore,

$$p_{uk} \cdot \mu_{sk} = p_{u2} \cdot \mu_{s2} = (0.2638) \cdot (239.0 \text{ ksi}) = 63.1 \text{ ksi}$$

$$p_{uk} \cdot (\sigma_{sk}^2 + \mu_{sk}^2) = p_{u2} \cdot (\sigma_{s2}^2 + \mu_{s2}^2) = 0.2638 \cdot ((239.0 \text{ ksi})^2 + (4.3 \text{ ksi})^2) = 15077.2 \text{ (ksi)}^2$$

Similar calculations to those above for Wire Group 2 are carried out for Wire Groups 3 and 4, as shown in Table 7-8.

Table 7-8 Calculations for mean tensile strength of the cable

Wire Group k	Number of Wires in Each Group N_k	Wires in Each Group w/out N_5 N_k	Fraction of Cable in Each Group p_{uk}	Tensile Strength Mean μ_{sk}	Tensile Strength Standard Deviation σ_{sk}	$p_{uk} * \mu_{sk}$	$p_{uk} (\mu_{sk}^2 + \sigma_{sk}^2)$
2	2274	2274	0.2639	239.0	4.3	63.1	15077.2
3	5263	5263	0.6107	235.9	5.7	144.1	34004.5
4	1081	1081	0.1254	231.1	8.7	29.0	6708.6
5	1360						
totals =>	9978		1.0000			236.1	55790.3
		$N_{eff} - N_5 =$ 8618					

Referencing Table 7-8, the mean tensile strength and standard deviation of the cable at this panel location is calculated as (note some differences in the values may exist due to rounding):

$$\mu_s = \sum_{k=2}^{4} (p_{uk} \cdot \mu_{sk}) = (63.1 \text{ ksi} + 144.1 \text{ ksi} + 29.0 \text{ ksi}) = 236.1 \text{ ksi}$$

$$\sigma_s = \sqrt{\left(\sum_{k=2}^{4} p_{uk}(\sigma_{sk}^2 + \mu_{sk}^2)\right) - \mu_s^2} = \sqrt{(55790.3 \text{ ksi}^2) - (236.1 \text{ ksi})^2} = 6.3 \text{ ksi}$$

In accordance with Article 5.3.3.1.2 of *NCHRP Report 534*, the cable strength for the Simplified Model is calculated using the equation:

$$R = (N_{eff} - N_5) \cdot a_w \cdot \mu_s \cdot K \qquad (5.3.3.1.2\text{-}1)$$

where: K = reduction factor (Figure 5.3.3.1.2-1 of *NCHRP Report 534*)
a_w = nominal area of one wire

The reduction factor, K, is determined from Figure 7-7 as a function of the coefficient of variation, which is calculated as σ_s/μ_s.

$\sigma_s / \mu_s = 6.3 \text{ ksi} / 236.1 \text{ ksi} = 0.027$

Therefore, from Figure 7-7, the strength reduction factor, K, is determined to be 0.91.

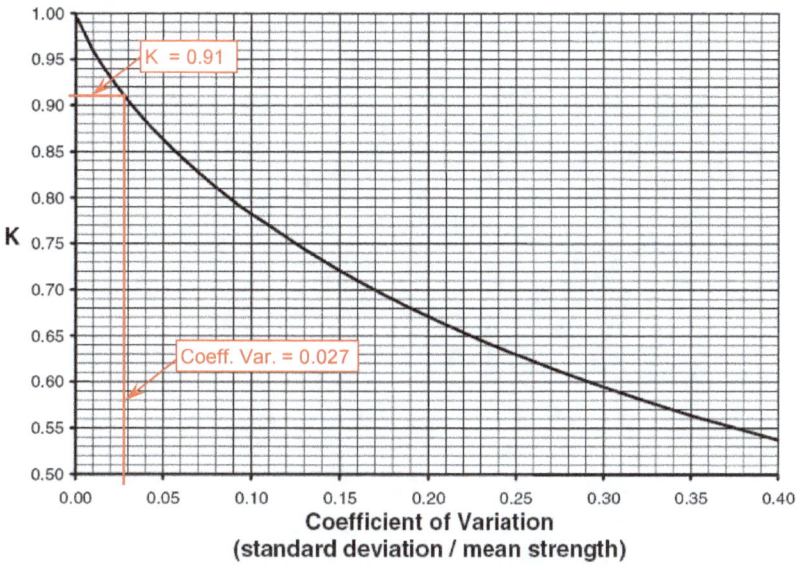

Figure 7-7 Graph for computing the strength reduction factor, K (taken from Figure 5.3.3.1.2-1 of *NCHRP Report 534*)

The estimated cable strength for Panel 77-78 is then calculated as:

$R = (8618 \text{ wires}) \cdot (0.0290 \text{ in}^2/\text{wire}) \cdot (236.1 \text{ ksi}) \cdot (0.91) = 53{,}683 \text{ kips}$

In a true investigation, this estimated cable strength for Panel 77-78, as calculated in accordance with the Simplified Model, can be compared with other panels that would be evaluated using the

Simplified Model. This will allow the engineer to determine which panel(s) would require further detailed calculations using the Brittle-Wire Model or the Limited Ductility Model.

8.0 APPENDIX B: PREVIOUS INSPECTION REFERENCES

Below is a list of references in which a suspension bridge cable has been inspected using the methodology presented in *NCHRP Report 534*.

1. Blasko, P.D., Borzok, M.J. and Kulicki, J.M., (2008), *Evaluating the Suspension Bridge Cables on North America's Busiest International Crossing*, Transportation Research Board (TRB) 2008 Annual Meeting CD-ROM, Washington, D.C.

2. Flint & Neill Partnership, (2006), *Audit of the Main Cable Inspection and Assessment, Final Report 1023-Rp-02 r4*, Project 34901MLNB, Forth Road Bridge, First Internal Cable Inspection, Scottish Executive, Gloucestershire, UK.

3. Mayrbaurl, R.M. and Goddard, H., (2007), *You Can't Judge a Cable by Its Cover*, Steel Bridge News, National Steel Bridge Alliance, Modern Steel Construction, pp. 43-47.

4. Benjamin Franklin Bridge, Philadelphia, Pennsylvania, using Brittle Wire Model; Weidlinger Associates Inc.

5. Walt Whitman Bridge, Philadelphia, Pennsylvania, using Brittle Wire Model; Weidlinger Associates Inc.

9.0 APPENDIX C: FLOWCHARTS

9.1 INSPECTION FLOWCHART

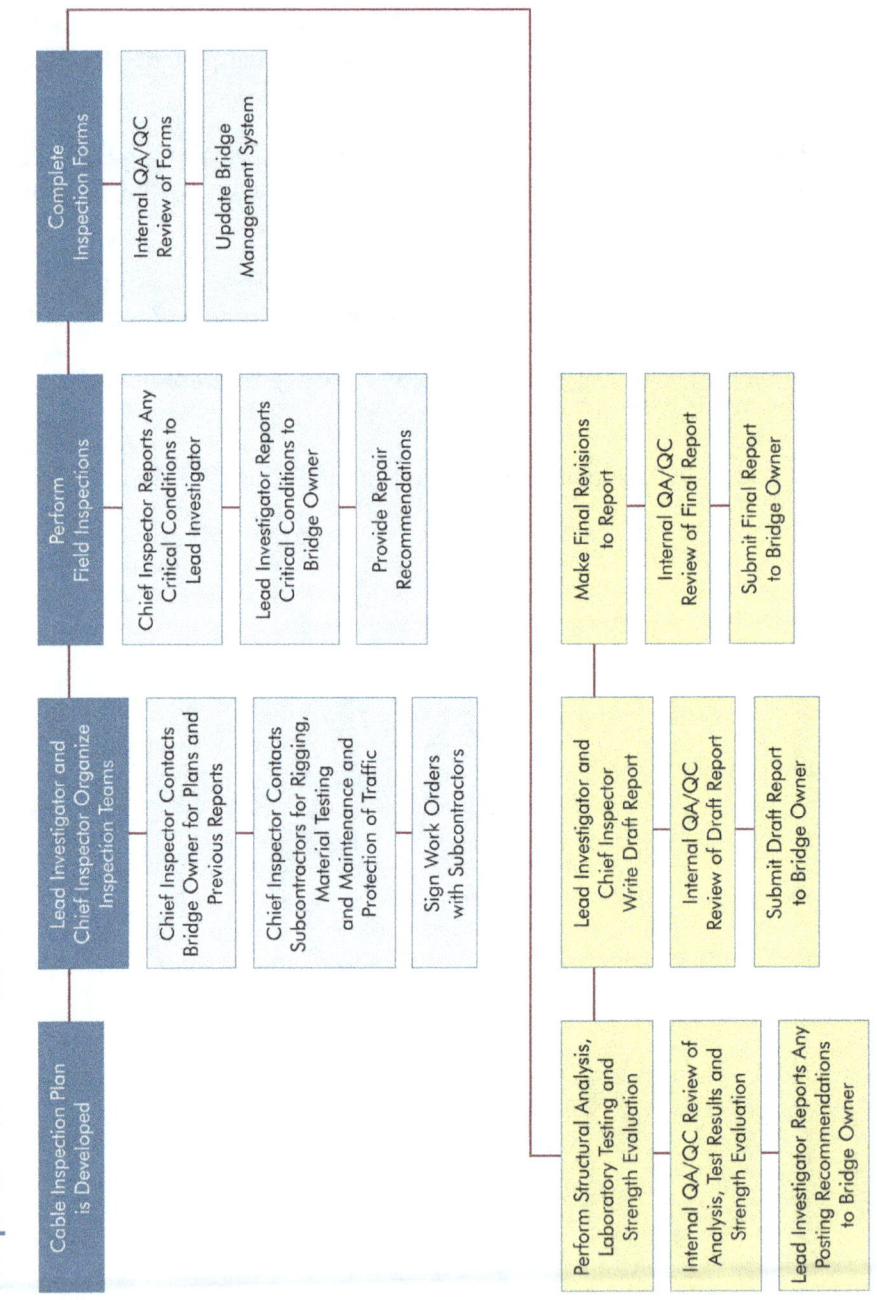

9.2 STRENGTH EVALUATION FLOWCHART

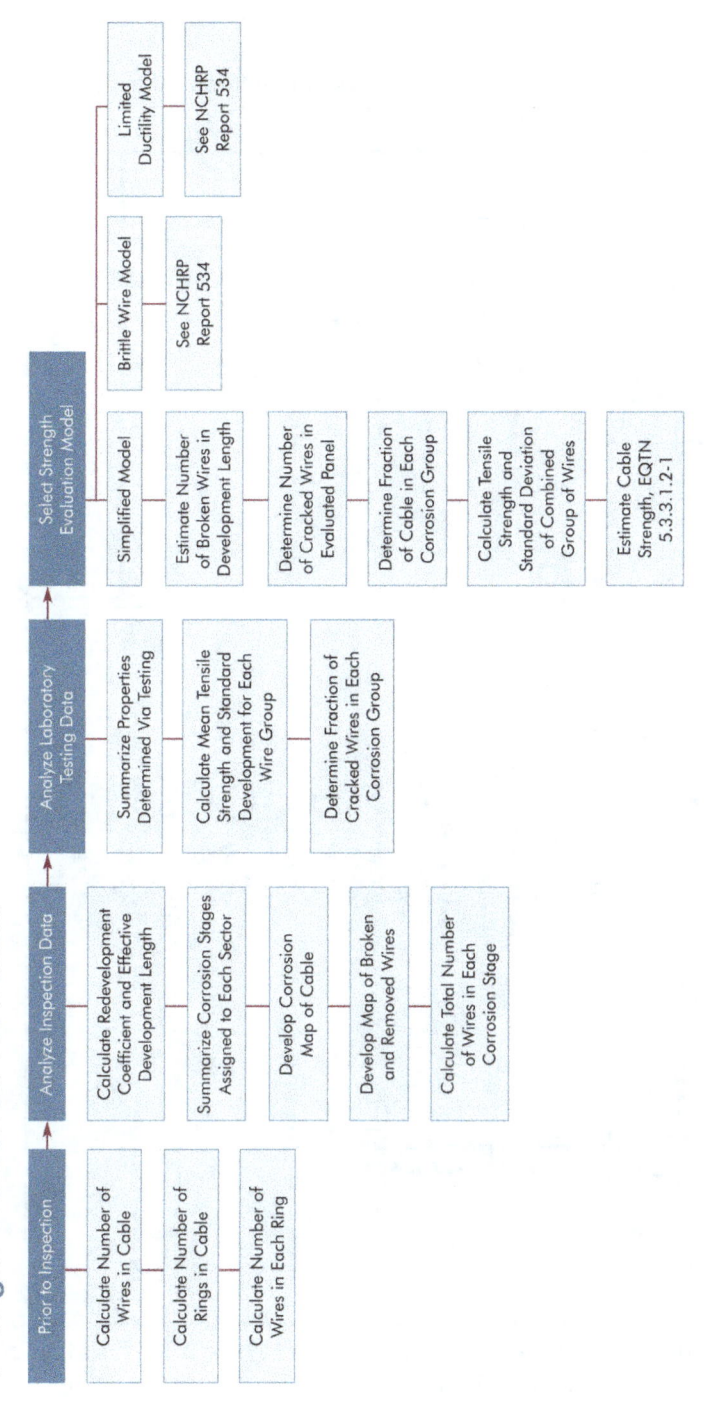

10.0 APPENDIX D: INSPECTION AND EVALUATION FORMS

10.1 INSPECTION FORMS

Inspection Forms, presented in Section 2 of *NCHRP Report 534*, are reproduced herein.

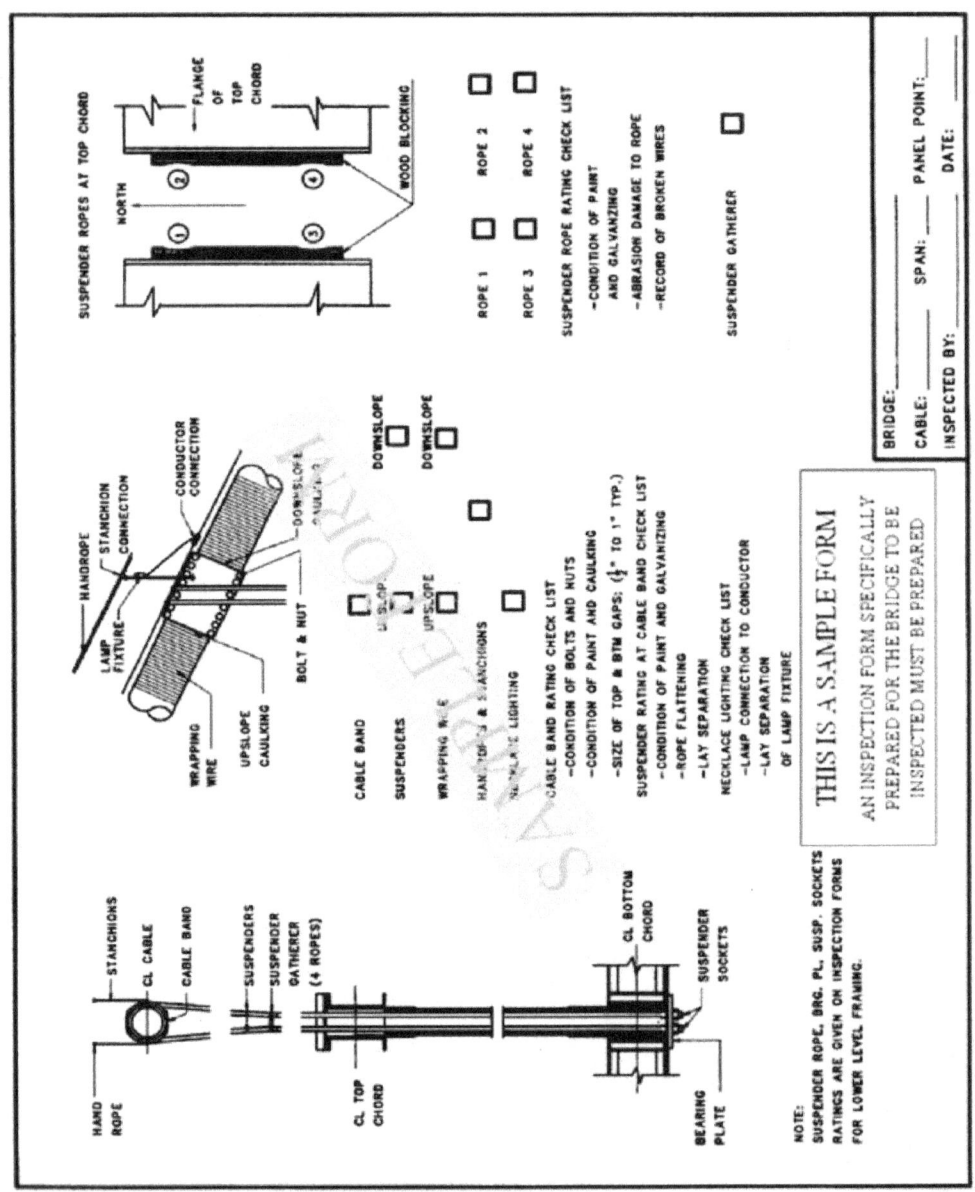

Figure 10-1 Typical cable biennial inspection form (taken from Figure 2.2.3.1-1 of *NCHRP Report 534*)

BD 188 (1/96)

BIN _____

NYS DEPT. OF TRANSPORTATION
BRIDGE INSPECTION REPORT

SHEET ____ OF ____

TEAM LEADER: _____ **ASST. TEAM LEADER:** _____ **DATE** _____

Feature Carried: _____
Feature Crossed: _____

REMARKS: TP 350 - [28] PRIMARY MEMBERS • CABLE GENERAL NOTES •

General Notes for Structural Condition Rating Criteria.

The cable system consists of four cables labeled "A", "B", "C" and "D" from South to North. Each cable is made up of 9,472 galvanized parallel wires and wrapped mainly with galvanized steel wires. There are a few locations near and inside the tower bell housing and near the saddles where the cables are wrapped with neoprene. The cables are in good condition.

The wrapping wire has failed in a few locations. The failed wrapping consists of either broken wires or loose wires with gaps. There are also sections where the wrapping wires are overlapping. The rating on the wrapping wire is not structural, but was lowered due to these deficiencies. The following is the rating criteria for the wrapping:

RATING	DESCRIPTION OF DEFECT
1	the wrapping wire is loose or missing in the entire panel
3	The wrapping wire section has loose or broken wires.
4	The wrapping wire section has overlapping wires.
5	The wrapping wire section has no defects.

Near the tower bell housing, there are depressions in the wrapping wires. This condition does not affect the ability of the wrapping to protect the main cables, however, the rating for wrapping wire sections with this condition have been lowered due to the same deficiencies mentioned above.

The cable bands are in good condition. No loose cable band bolts or slippage of the bands along the cable are present. The caulking on several bands is loose or missing. This condition does not affect the rating of the main cable, but the rating of bands with these conditions have been lowered. The criteria for rating the cable bands are as follows.

RATING	DESCRIPTION OF DEFECT
1	All cable band bolts are loose or the band has slipped
3	All the caulking is missing or up to 20% of the bolts are loose
4	Up to 25% of the caulking is missing
5	The cable band has no defects.

The structural rating of the main cables are not affected by the deficiencies in the wrapping wire or the cable bands. Therefore, the main cables are rated 5.

100% HANDS ON INSPECTION WAS PERFORMED ON ALL NON-REDUNDANT CABLES.

Figure 10-2 Typical summary form showing biennial inspection rating system (taken from Figure 2.2.3.1-2 of *NCHRP Report 534*)

BD 188 (1/96)							
BIN _____						NYS DEPT. OF TRANSPORTATION BRIDGE INSPECTION REPORT	
						SHEET ____ OF ____	
TEAM LEADER: _____ L				ASST. TEAM EADER: _____		DATE _____	
Feature Carried: _____							
Feature Crossed: _____							

				TP 350 - [28] PRIMARY MEMBERS			
RATINGS			PHOTO NO.	CABLE A			
NEW	PREV	PAINT		LOC. & SPAN	PP	MEMBER	REMARKS
5	5	5		BMS/10	81	band	
5	5	3		BMS/10	81-82	wrap	
5	5	5		BMS/10	82	band	
5	5	4		MMS/10	81-82	wrap	
5	5	5		MMS/10	81	band	
5	5	3		MMS/10	80-81	wrap	
5	5	5		MMS/10	80	band	
4	4	3	89S	MMS/10	79-80	wrap	There are overlapping wires near PP 80.
4	5	5		MMS/10	79	band	The seal between the cable band and the bottom portion of the cable is missing.
5	5	3		MMS/10	78-79	wrap	
5	5	5		MMS/10	78	band	
5	5	3		MMS/10	77-78	wrap	
4	5	5		MMS/10	77	band	The seal between the cable band and the bottom portion of the cable is missing.
4	4	3	89S	MMS/10	76-77	wrap	There are overlapping wires near PP 80.
4	5	5		MMS/10	76	band	The seal between the cable band and the bottom portion of the cable is missing.
5	5	3		MMS/10	75-76	wrap	
5	5	5		MMS/10	75	band	
5	5	3		MMS/10	74-75	wrap	
5	5	5		MMS/10	74	band	
5	5	3		MMS/10	73-74	wrap	
5	5	5		MMS/10	73	band	
5	5	3		MMS/10	72-73	wrap	
5	5	5		MMS/10	72	band	
5	5	3		MMS/10	71-72	wrap	
5	5	5		MMS/10	71	band	
5	5	3		MMS/10	70-71	wrap	
5	5	5		MMS/10	70	band	
5	5	3		MMS/10	69-70	wrap	
5	5	5		MMS/10	69	band	
5	5	3		MMS/10	68-69	wrap	
5	5	5		MMS/10	68	band	
5	5	5		MMS/10	67-68	wrap	
5	5	5		MMS/10	67	band	

Figure 10-3 Typical form for biennial inspection showing detailed ratings (taken from Figure 2.2.3.1-3 of *NCHRP Report 534*)

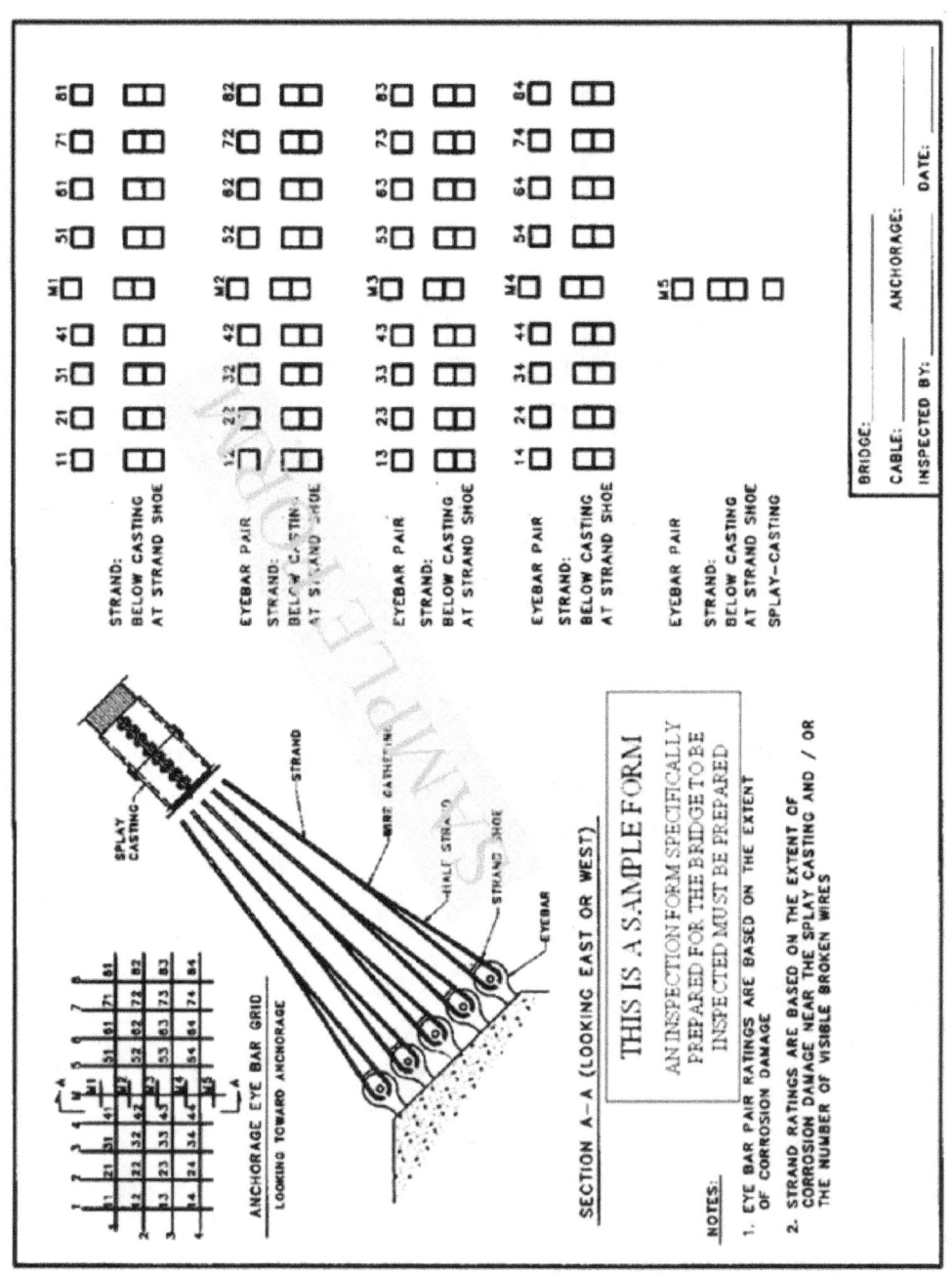

Figure 10-4 Typical form for biennial inspection of cable inside anchorage (taken from Figure 2.2.3.2-1 of *NCHRP Report 534*)

Authority ______________

Bridge ______________

TYPICAL CABLE INSPECTION LOCATIONS DIAGRAM

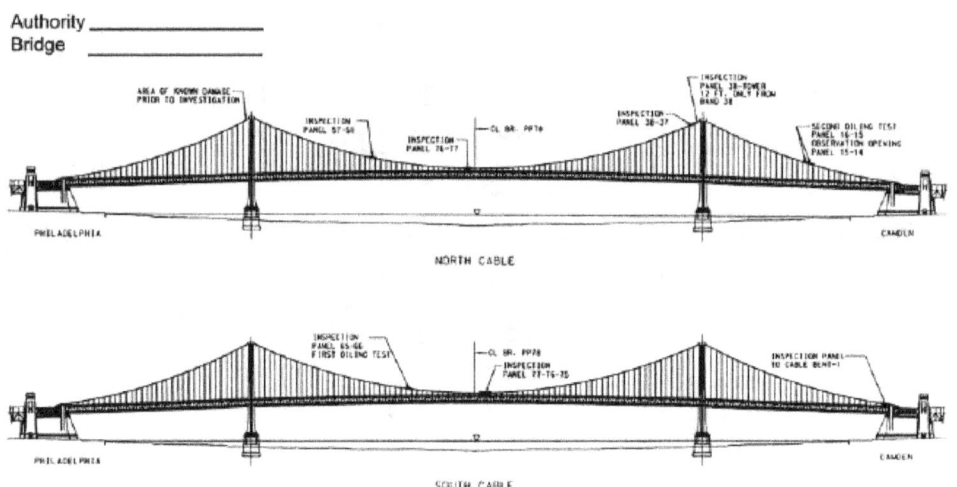

Figure 10-5 Form for recording locations of internal cable inspections (taken from Figure 2.2.1.2.4-1 of *NCHRP Report 534*)

Figure 2.3.1.2.4-1. Form for recording locations of internal cable inspections.

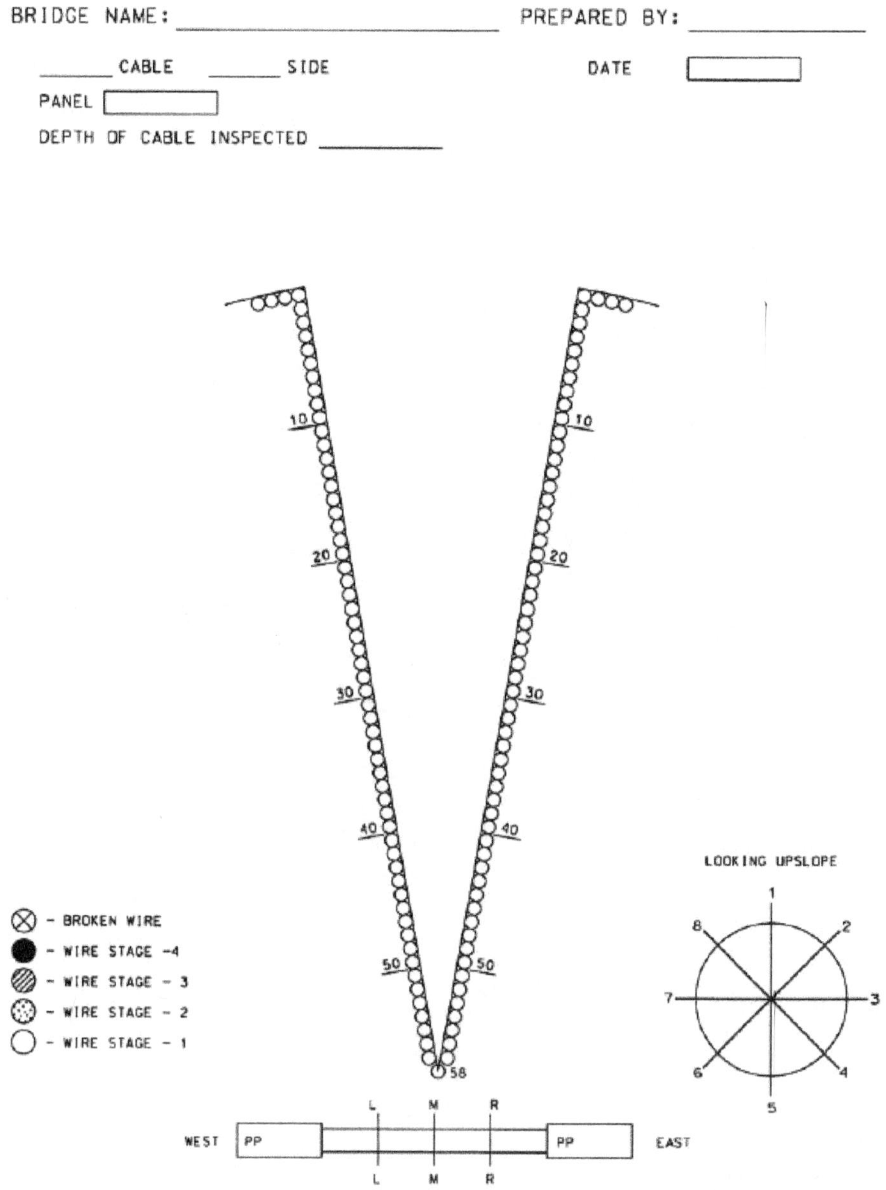

Figure 10-6 Form for recording observed wire damage inside wedged opening (taken from Figure 2.3.1.2.4-2 of *NCHRP Report 534*)

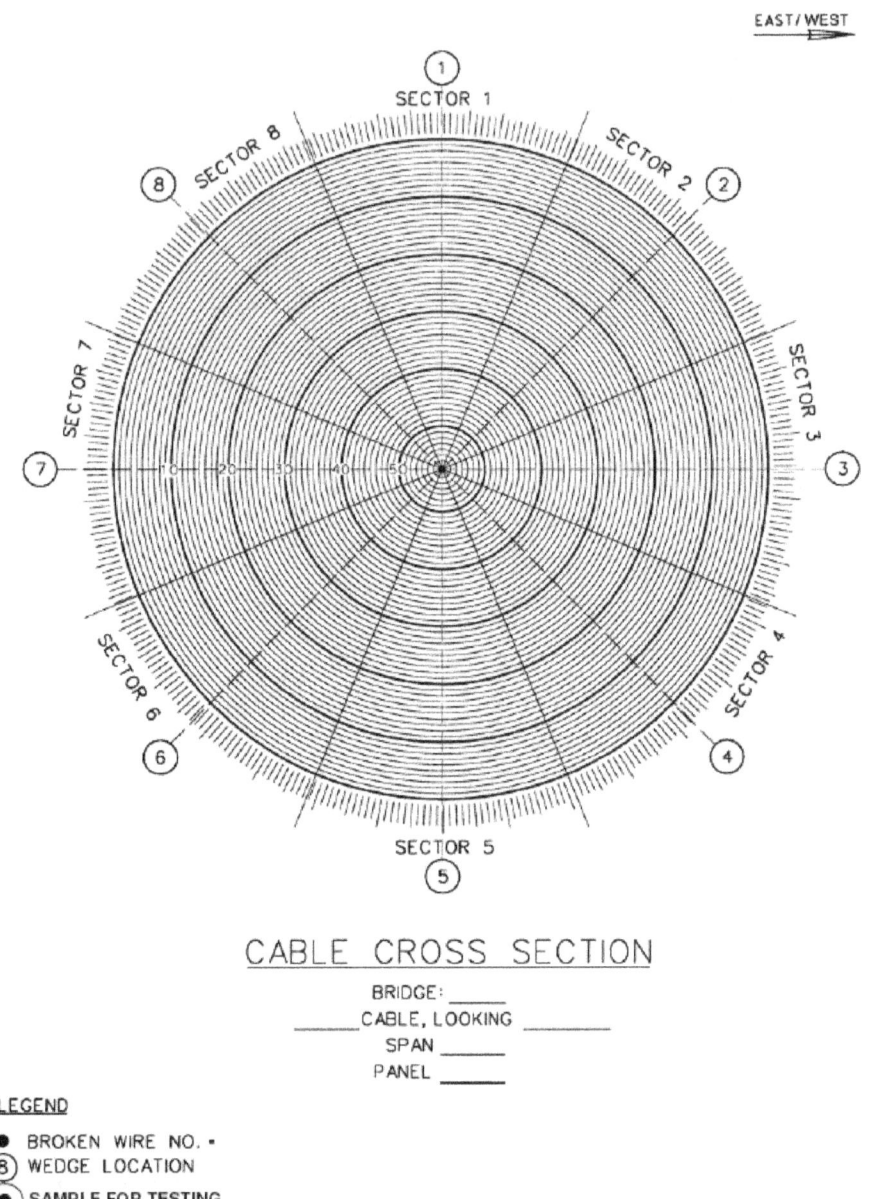

Figure 10-7 Form for recording locations of broke wires and samples for testing (taken from Figure 2.3.1.2.4-3 of *NCHRP Report 534*)

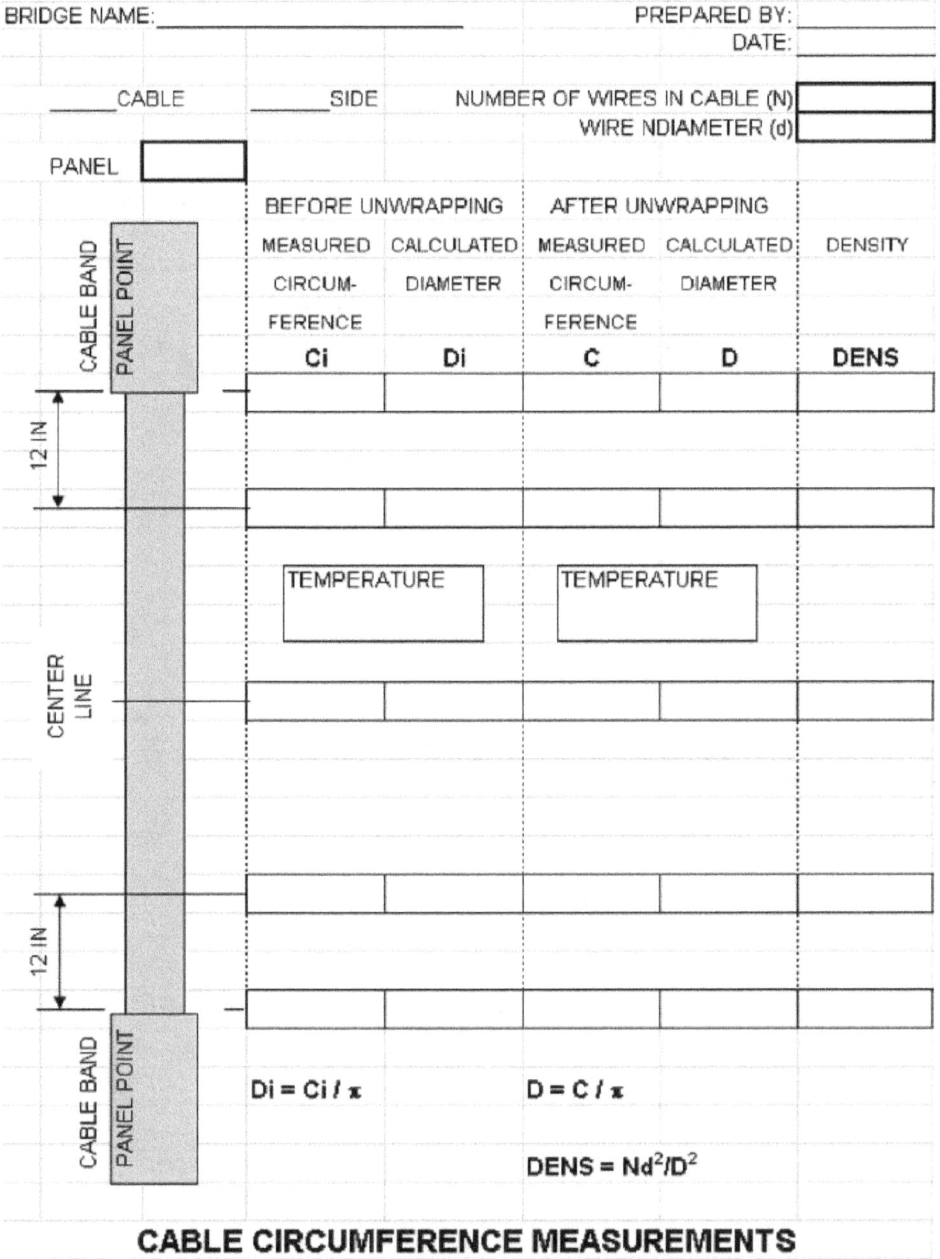

Figure 10-8 Form for recording cable circumference (taken from Figure 2.4.1-1 of *NCHRP Report 534*)

10.2 STRENGTH EVALUATION FORMS

Tables used for summaries and calculations associated with the Strength Evaluation example shown in Appendix A are shown in this section. The tables are presented so that the reader can use them or develop similar tables for a strength evaluation.

Table 10-1 Typical table that can be used for assignment of corrosion stages to individual rings of wires

Ring Number, d	Number of Wires in Ring, n	GRADING OF DETERIORATED WIRES SECTOR NUMBER														
		1		2		3		4		5		6		7		8
		L	R	L	R	L	R	L	R	L	R	L	R	L	R	L	R
1	n_x																
2	$n_{(x-1)}$																
3	$n_{(x-2)}$																
4	$n_{(x-3)}$																
5	$n_{(x-4)}$																
6	$n_{(x-5)}$																
7	$n_{(x-6)}$																
8	$n_{(x-7)}$																
9	$n_{(x-8)}$																
10	$n_{(x-9)}$																
11	$n_{(x-10)}$																
12	$n_{(x-11)}$																
13	$n_{(x-12)}$																
14	$n_{(x-13)}$																
15	$n_{(x-14)}$																
16	$n_{(x-15)}$																
17	$n_{(x-16)}$																
18	$n_{(x-17)}$																
19	$n_{(x-18)}$																
20	$n_{(x-19)}$																
⋮																	
Center Ring	n_1																

Table 10-2 Typical table that can be used for assignment of corrosion stages to half sectors of cable

Ring Number, d	Number of Wires in Ring, n	Number of Half-Sectors in Each Corrosion Stage Stage Number, k				Number of Wires in Each Corrosion Stage, N_{sk} Stage Number, k			
		1	2	3	4	1	2	3	4
1	n_x								
2	$n_{(x-1)}$								
3	$n_{(x-2)}$								
4	$n_{(x-3)}$								
5	$n_{(x-4)}$								
6	$n_{(x-5)}$								
7	$n_{(x-6)}$								
8	$n_{(x-7)}$								
9	$n_{(x-8)}$								
10	$n_{(x-9)}$								
11	$n_{(x-10)}$								
12	$n_{(x-11)}$								
13	$n_{(x-12)}$								
14	$n_{(x-13)}$								
15	$n_{(x-14)}$								
16	$n_{(x-15)}$								
17	$n_{(x-16)}$								
18	$n_{(x-17)}$								
19	$n_{(x-18)}$								
20	$n_{(x-19)}$								
⋮									
Center Ring	n_1								

Table 10-3 Typical table that can be used to record tension test results for a single wire

CABLE AND PANEL =			Stresses Based on:		
WIRE SAMPLE =			Diameter of wire =		in
			Area of wire =		in^2

CABLE AND PANEL DESIGNATIONS		LEGEND
W = WEST	M = MAIN	X = NUMBER OF SAMPLE WIRE
E = EAST	S = SIDE	X.01 = SPECIMEN No. 1 FROM WIRE X
N = NORTH CABLE	2123 = PANEL 21-23	X.02 = SPECIMEN No. 2 FROM WIRE X
S = SOUTH CABLE	ANCH = ANCHORAGE	X.91 = LONG SPECIMEN FROM WIRE X

Specimen Number	Corrosion Stage	Max Load (lbs)	Yield Strength (ksi)	Tensile Strength (ksi)	Elongation in 10" (%)	Reduction in Area (%)	Remarks	Fracture Type
XX.01								
XX.02								
XX.03								
XX.04								
XX.05								
XX.06								
XX.07								
XX.08								
XX.09								
XX.10								
XX.11								

11 = number of samples

Mean
Standard Deviation
Maximum
Minimum

FRACTURE TYPES
- A — Ductile; Cup and cone
- B — Ductile; Cup and cone with shear lips alternating above and below fracture plane
- B-C — Semi-ductile; Ragged with partial shear lips and reduced reduction in area
- C — Brittle; Ragged with minimal or no reduction in area
- D — Brittle w/Crack; Fracture with partial crack

NOTE 1: Surface of wire covered with a gummy material, possibly dried linseed oil
NOTE 2: Surface corrosion is present at the fracture location, which is the probable initiation point of the fracture

L = LOCAL M = MODERATE S = SEVERE
O = OVERALL H = HEAVY

Table 10-4 Typical table that can be used to summarize the tension test results

CABLE AND PANEL =	
WIRE SAMPLE =	
YEAR =	

Stresses Based on:		
Diameter of wire =		in
Area of wire =		in²

CABLE AND PANEL DESIGNATIONS
W = WEST   M = MAIN
E = EAST   S = SIDE
N = NORTH CABLE   2123 = PANEL 21-23
S = SOUTH CABLE   ANCH = ANCHORAGE

Cable and Panel	Wire Sample Number	Corr. Stage	Wire Group	Max Load (lbs)	Tensile Strength (ksi)	Remarks	Fracture Type	STRENGTH BY GROUP				
								1	2	3	4	5

XX = number of samples

Mean					
Standard Deviation					
GROUP Mean					
GROUP Standard Deviation					

FRACTURE TYPES
A	Ductile; Cup and cone
B	Ductile; Cup and cone with shear lips alternating above and below fracture plane
B-C	Semi-ductile; Ragged with partial shear lips and reduced reduction in area
C	Brittle; Ragged with minimal or no reduction in area
D	Brittle w/Crack; Fracture with partial crack

Table 10-5 Typical table that can be used to summarize the number of broken and cracked wires

Corrosion Stage k	Number of Wires in Each Stage N_{sk}	Net Number of Broken Wires $N_b - N_r$	Number of Unbroken Wires N_{0k}	Fraction of Unbroken Wires p_{ok}	Fraction of Cracked Wires $p_{c,k}$	Cracked Wires in Evaluated Panel $N_{c,k}$
2						
3						
4						
5						
totals =>						

$N = $
$N_{eff} = $
$N_5 = \Sigma N_{c,k} = $

Table 10-6 Typical table that can be used for calculating the mean tensile strength

Wire Group k	Number of Wires in Each Group N_k	Wires in Each Group w/out N_5 N_k	Fraction of Cable in Each Group p_{uk}	Tensile Strength Mean μ_{sk}	Standard Deviation σ_{sk}	$p_{uk} * \mu_{sk}$	$p_{uk}(\mu_{sk}^2 + \sigma_{sk}^2)$
2							
3							
4							
5							
totals =>							

$N_{eff} - N_5 = $

11.0 APPENDIX E: BTC METHOD FOR EVALUATION OF REMAINING STRENGTH AND SERVICE LIFE OF BRIDGE CABLES[2]

11.1 INTRODUCTION

Given the importance of cable-supported bridges, random sampling and reliability-based analytical techniques are required for the assessment of remaining cable strength. State-of-the-art assessment techniques employ reliability criteria (similar to LRFD criteria), in which strength, strain and loads are known as probabilistic quantities. If an evaluation is conducted using these criteria, the results can be used to establish the tempo of bridge cable inspection and further evaluations in the future. Once probability distributions for wire mechanical properties, such as strength and strain, and loads are established, it's possible to develop a cable failure mechanism and assess the serviceability of the cable. Use of probabilistic analysis in this approach is similar to the LRFD probabilistic analysis employed in the current AASHTO standards.

Conventional cable strength evaluation methods rely heavily on the visual assessment of surface corrosion observed on individual wire surface. Wire sampling and strength evaluation, employed in these conventional methods, are dependent on visual condition of wires. This Appendix presents the BTC method, a probability-based, and U.S. patented [E.1][3] methodology for the evaluation of remaining strength and service life of both; parallel wire and helical wire bridge cables. The method applies to both zinc-coated and bright bridge wire. The method includes random sampling of individual cable wires from each investigated panel, mechanically testing the sampled wires, determining the probability of broken and cracked wires, evaluating the ultimate strength of cracked wires using fracture toughness criteria and utilizing the above data to assess the remaining strength of the cable in each panel. The probabilistic-based method assesses remaining service life of the cable by determining the rate of change of broken and cracked wires detected over a time frame, measuring the rate of change in effective fracture toughness over same time frame, and applying the rates of change to a time-dependent strength degradation prediction model. The BTC method provides sensitivity analysis to identify the key inputs which influence the estimated cable strength and assist the bridge owner in the decision making process. As such, the BTC method provides the bridge owner a greater understanding of the deterioration mechanisms at work in the cables as well as a comprehensive methodology that results in a higher level of confidence in the estimated strength and assessed remaining life of the bridge cable. The BTC method has been peer-reviewed by MTA Bridges & Tunnels, New York State Bridge Authority and New York State Department of Transportation. To date, the BTC method has been applied to assess remaining strength and residual life of the main suspension cables at the Bronx-Whitestone Bridge in New York City, and the Mid-Hudson Bridge, in Poughkeepsie, in the state of New York, USA.

---

[2] Appendix E is authored solely by Khaled M. Mahmoud, Ph.D., P.E., Bridge Technology Consulting, New York City, email: khaled@kmbtc.com. Persons interested in using the BTC Method should contact the author of this Appendix. The inclusion of Appendix E does not necessarily constitute an endorsement of the BTC Method by the FHWA or the authors of this Primer.

[3] Numbers in brackets refer to References provided at the end of this Appendix.

This Appendix provides a brief description of the BTC method for evaluation of remaining strength and residual life of bridge cables.

11.2 MAIN CABLE INSPECTION AND SAMPLING

Internal inspection of main cables is recommended at a number of panels on each cable, depending on length of main cable, and known condition from history of previous cable investigations. For cables that never received internal inspections, a four to six panels are to be wedged, internally inspected and sampled from on each main cable. Prior to field inspection, a random sampling plan is prepared to extract wires for testing from each investigated panel.

11.2.1 Panel Selection Criterion

The main goal of in-depth cable inspection is to assess the damage in the most deteriorated panels. Assessment of the structural integrity of the cable is achieved by calculating the factor of safety for each of the investigated panels. The panel with the lowest factor of safety will govern the factor of safety for the entire cable. Thus, our objective is to choose the most *at-risk* panels, and not random panels to calculate the safety factor for the cable.

To that end, a study is conducted of the recorded history of external and internal cable inspections, wire breaks, and wire tests along the cable length to identify the most *at-risk* panels. Based on this history, a selection is made of the panels to be opened for internal inspection and sampling of wires. For main cables that have no previous history of internal inspection and show no evident external signs of deterioration, panels are randomly selected.

11.2.2 Random Sampling and Sample Size Determination

Because it is unfeasible to sample and test every wire in the cable, only a sample of wires is removed for testing in the laboratory. In this way, sampling is done to generate a small group of wires that is as similar to the entire population of wires as possible. With that in mind, two questions arise; how well does the sample represent the larger population from which it was drawn? How closely do the features of the sample resemble those of the larger population? To answer these questions; a definition of sampling methods is introduced first. Sampling methods are classified as either *probability* or *nonprobability*. In probability samples, each member of the population has a known non-zero probability of being selected. Probability methods include random sampling, systematic sampling, and stratified sampling. In nonprobability sampling, members are selected from the population in some nonrandom manner. These include convenience sampling, judgment sampling, quota sampling, and snowball sampling. The advantage of probability sampling is that sampling error can be calculated. Sampling error is the degree to which a sample might differ from the population. When inferring to the population, results are reported plus or minus the sampling error. In nonprobability sampling, the degree to which samples might statistically differ from the population remains unknown. Thus random sampling presents the best representation and resemblance of wire condition throughout the entire bridge cable.

11.2.2.1 Random Sampling and Practical Considerations

In random sampling, each wire in the available pool of wire samples has an equal and known chance of being selected. Random sampling procedures do not guarantee that the sample is representative, but they do increase the probability that the randomly selected wires will be representative of cable condition. There is a sampling error in the estimated cable strength because not every wire is sampled and tested. The sampling error describes the range that the estimated cable strength is likely to fall within.

Sampling should be limited to provide an acceptable level of error in the estimated cable strength. This is to minimize vulnerabilities introduced in the cable cross section due to the sampling and removal of wires. A given cable investigation is not the last opportunity that wire samples will be removed from cable. Therefore, the following practical considerations must be recognized:
- It is not feasible to remove wires too deep in the wedge opening due to clearance problems in the wedge opening when cutting, splicing and re-tightening the splice.
- Even if a deeper wire is pulled with the use of a special tool out of the wedge, access for splicing and re-tightening would be very limited and damage to zinc coating of neighboring wires would become more likely.
- It is therefore our recommendation to minimize damage to main cable and not to remove samples from areas where they could not effectively be replaced and spliced.
- Outer wires are easily accessible, however, inner wires are difficult to reach for the purpose of tightening ferrules, and often a wire would be spliced with zero or small stress. This is evident by the slack condition of many spliced wires observed on suspension bridge cables.

In the random sampling plan, we define the *sampling frame* as the accessible group of wires that samples will be randomly selected from. Sampled wires constitute the sample size from which valid conclusions about the entire cable are based. This statistical inference is done with the aim of inferring the degraded condition of all wires from those found in the observed sample. By virtue of random selection of sampled wires, different conditions of wires would be included in the sample.

11.2.2.2 Sampling Size Determination

It is important to evaluate the effect the sample size has on the error in the estimated cable strength. The error results from imprecision associated with estimating nominal cable strength based on a limited number of wire samples. Therefore an acceptable target level of error is set for the estimated cable strength. According to Article 3.4 of NCHRP Web Document 28 [E.2], the minimum expected finite fatigue life is taken as the fatigue resistance two standard deviations below the mean fatigue resistance. This is equivalent to a 97.73% one-sided confidence level. For the purpose of determining the sample size, we assume that the cable strength is determined only by the ultimate strength. The ultimate strength data, from previous investigations, if it exists, is fitted to an appropriate probability distribution. The sample size is then determined based on a correlation between the strength distribution parameters and different level of error at a given level of confidence.

In the case where no previous history of testing exists, sample size determination could be guided by random sample size established at other bridges by the BTC method.

11.2.2.3 Wedge Pattern

Per the BTC method, the wedge pattern could be selected randomly. The eight-wedge pattern, which is shown in Figure 11-1 for illustrative purposes, is typically used for internal inspection of main suspension cables.

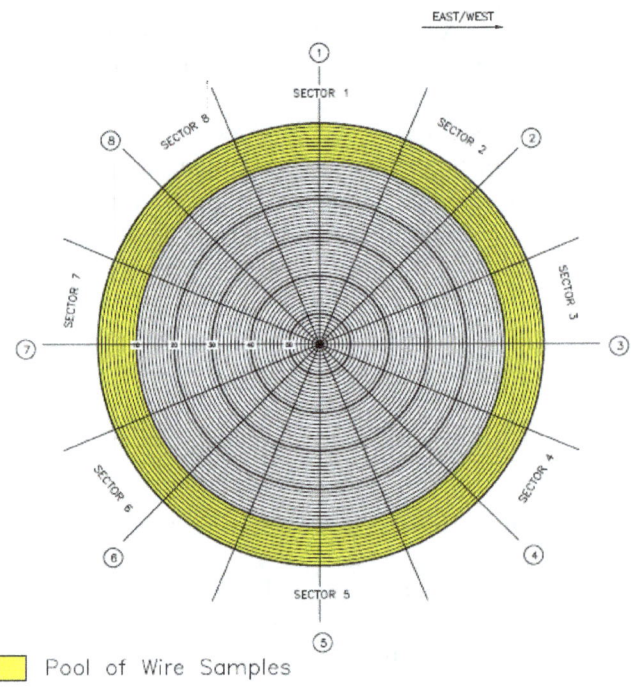

Figure 11-1 Eight-wedge pattern and pool of wire samples in the cable cross section

11.2.2.4 Sampling Frame of Random Sample

The sampling frame, which is the pool of wire samples, should be limited to the group of wires that can be accessed for cutting, and splicing back to service load. This *sampling frame* depends on the diameter of the cable and length of panel opening, where larger diameter cables and longer panel opening provide more space for splicing sampled wires deeper in the wedge opening. For demonstration purposes, Figure 11-1 shows sampling frame, for a cable with 16-inch diameter and 6,080 wires. The first ten (10) rings of wires define the sampling frame; i.e. to a depth of approximately 2-inches which totals 160 wires. Those 160 wires are deemed easily accessible for cutting, splicing and retightening back to service load. During the course of inspection, each wire that is randomly selected is assigned an I.D. number.

Figure 11-2 shows a tagged wire with I.D. that reads PP 3-4S, W7/8, R4. This identifies a wire sampled from P.P. 3-4, in Wedge #7, Wedge #8 side, Ring # 4.

Figure 11-2 Photo showing the sample tag in wedge #7, wedge #8 side, ring #4, panel point 3-4

11.2.3 Inspection Procedures

An access platform is typically constructed along the full length of each of the panels subject to internal inspection. After the removal of the Gauge 9 wrapping wire in each panel, the cable is typically wedged along four planes with an eight wedge pattern as shown in Figure 11-1, numbered in sectors from 1 to 8.

The cable is wedged by the Contractor, utilizing plastic wedges driven to the center of the cable, as shown in Figure 11-3. As the photo shows, a stack of wedges was driven to allow for proper inspection and access to splice sampled and broken wires. The BTC method identifies proportion of broken wires based on wires found broken during inspection in each panel. The proportion of broken wires is treated as a probabilistic quantity. Figure 11-3 shows an interior broken wire that extended out of wedge opening following the driving of wedges. Broken wires are spliced to sustain service load if they are accessible for splicing.

Figure 11-3 Photo showing wedges driven to the center of the cable, interior broken wire is shown

Inspectors determine the visual condition of corrosion stages based on visual assessment of corrosion condition of the wire surface into four stages of corrosion that were first introduced by Hopwood and Havens in 1984 [E.3], [E.4], [E.5]. The evaluation of remaining strength per the BTC method is independent of the visual evaluation of stages of corrosion. However the four stages of corrosion are defined, as follows, for illustration of the inspection procedure, see Figure 11-4.

Stage 1: the zinc coating of wires is oxidized to form zinc hydroxide, known as "white rust".
Stage 2: the wire surface is completely covered by white rust.
Stage 3: appearance of a small amount of (20-30% of wire surface area) of ferrous corrosion due to broken zinc coating.
Stage 4: the wire surface is completely covered with ferrous corrosion.

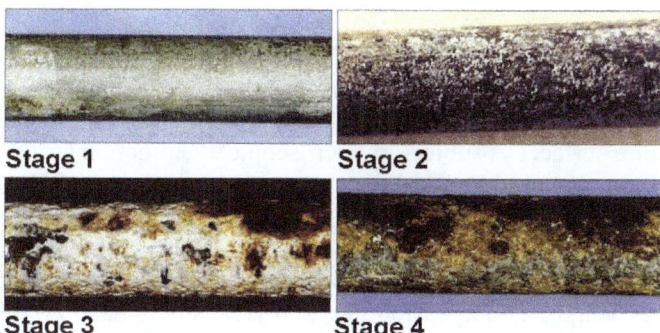

Figure 11-4 Photographs showing the four stages of corrosion

11.3 WIRE TESTING PROGRAM

Per the BTC method, the following tests are performed on the sample of wires collected from the bridge cable during internal inspection. Standard (18 inch) and long (60 inch to 72 inch) wire specimens are taken from each wire sample for testing.

11.3.1 Enhanced Tensile Strength Test in Standard Wire Specimens

Each standard (18-inch) specimen is subject to enhanced tension test in which the full range of the stress strain curve is recorded. Mechanical properties of wires are measured by conducting tensile tests. A gage LVDT type extensometer is used for measurement of strain; to record true strain up to failure. Full stress strain curve for each specimen is provided with the measured ultimate strength, ultimate elongation, as measured along a gage length, Young's Modulus, and yield strength. These properties of standard specimens are used for the evaluation of remaining cable strength. Fracture surfaces of all tested wire specimens are examined for presence of preexisting cracks.

11.3.2 Tensile Strength Test on Long Wire Specimens

Each long (60-inch to 72-inch) specimen is subject to tension test, where length of the long specimen is determined based on maximum available distance between grips of the testing machine. The purpose of testing long specimens is to improve our ability in detection of preexisting cracks in wire test specimens. The probability of finding cracks in long specimens is higher than that of locating cracks in standard specimens. The reason for that is that in standard specimens, only about 12-inch of the specimen length is outside the grips of the testing machine. Therefore cracks might be missed if the tension test is limited only to standard specimens. For each long specimen, the measured ultimate strength, crack location and depth are provided.

11.3.3 Fracture Toughness Test

Fracture toughness tests are performed on a group of wires from different panels for evaluation of fracture toughness of high strength cable wire [E.1]. The results of fracture toughness test are utilized in the assessment of remaining cable strength and residual life of the cable.

11.3.4 Fractographic and Scanning Electron Microscope (SEM) Evaluation

Following the enhanced tension test on standard specimens, and tension test on long specimens, the entire set of fracture surfaces is subject to microscopic examination to identify preexisting cracks. The fractures are photographed at different magnifications (e.g., 20X, 100X, and 600X) to show the morphology of fracture surface, and crack depth for each cracked specimen. It is important to emphasize that the full stress strain curve for each cracked specimen shall be provided by the testing laboratory with the measured ultimate strength, ultimate elongation, as measured along a gage length, Young's Modulus, and yield strength, and fracture strength for each test specimen. Measured crack depth and measured fracture toughness are used to establish the average ultimate strength of cracked wire proportion, as shown later.

Scanning electron microscope (SEM) evaluation is provided for select test specimens with different crack depths to illustrate mode of crack propagation and morphology of fracture surface.

11.4 CABLE STRENGTH EVALUATION

In the BTC method, the cable is represented as a series of parallel wires. The wire fails when the weakest segment fails. Therefore, the wire is as strong as its weakest segment. In ASTM standard tension test, the wire demonstrates almost perfect elastic behavior until plastification occurs at a randomly located weakest section. Incremental increase in strain usually leads to strain hardening in the weakest section, allowing for further load increase associated with deformation concentration until ultimate strength is reached. This behavior displays elastic-plastic behavior of steel, where the specific characteristics depend on the manufacturing process of the wire.

11.4.1 BTC Method Inputs

The stress strain curves of wire specimens tested in the lab display a linear relationship between stress and strain up to the yield point and a nonlinear behavior between the yield and the ultimate point. This behavior is described by the Ramberg-Osgood relationship as follows:

$$\varepsilon = \frac{\sigma}{E} + \alpha\, \varepsilon_e \left(\frac{\sigma}{\sigma_e}\right)^n \qquad (11\text{-}1)$$

where σ, ε, E, are the stress, strain, and Young's modulus respectively, n is the strain hardening exponents, ε_e, is the yield strain, σ_e is the yield strength and α is a fitting parameter.

The test data provides measurements for yield strain, ε_o, Young's Modulus, E, ultimate strength, σ_u, and ultimate strain, ε_u. Numerous test data of deteriorated wires show a larger scatter in the ultimate strain, compared with that of the ultimate strength, where coefficient of variation for ultimate strain is multiples of the coefficient of variation for ultimate strength. The coefficient of variation, for a given variable, is obtained by dividing its standard deviation by its mean value.

The noted significant variation in the ultimate strain illustrates the importance of including ultimate strain along with ultimate strength data in evaluation of remaining cable strength. The BTC method uses, as input, the four variables yield strain, ε_o, Young's Modulus, E, ultimate strength, σ_u, and ultimate strain, ε_u, where each variable is described by an appropriate probability distribution, and the cable strength is estimated separately for each investigated panel. In each panel, the inputs include proportions of broken and cracked wires, which are treated as probabilistic quantities. The fracture toughness along with crack depth information are used to determine the average ultimate strength of cracked wires, $\sigma_{ult\ cr.}$, as follows:

$$\sigma_{ult\ cr} = \frac{K_c}{Y\left(\dfrac{a}{D}\right)\sqrt{\pi a_c}} \tag{11-2}$$

where K_c is the effective fracture toughness, a_c is the critical crack depth and $Y\left(\dfrac{a}{D}\right)$ is the crack geometry factor.

11.4.2 Choice of Probability Distributions

According to the Engineering Statistics Handbook [E.6], life distribution models are chosen for one or more of the following three reasons:
- There is a physical/statistical argument that theoretically matches a failure mechanism to a life distribution model
- A particular model has previously been used successfully for the same or a similar failure mechanism
- A convenient model provides a good empirical fit to all the failure data.

Whatever method is used to choose a model, the model should "make sense". Distribution models such as the lognormal and the Weibull are so flexible that it is not uncommon for both to fit a small set of failure data. Since data for these four variables is used to infer the estimated cable strength, it is important to "test" whether the distributions chosen are consistent with the collected data. For goodness of fit, different statistical tests are used; such as Anderson-Darling, Chi-Square, or Kolmogorov-Smirnov, to decide whether a distribution model under examination is acceptable.

11.4.3 Elongation Threshold Criterion, $M_{threshold}$

The ultimate elongation of wire is utilized in the BTC method to classify wires into two groups. Wires that fail at an ultimate elongation lower than a specific threshold elongation, $M_{threshold}$, are categorized under the *worst-wire proportion*. All other wires, i.e., wires that demonstrate higher elongation than the threshold elongation, $M_{threshold}$ are classified as the *better-wire proportion*. By the above definition, the *worst-wire proportion* contains all cracked and broken wires, as well as some intact wires, while the *better-wire proportion* contains only intact wires.

To establish a threshold elongation, $M_{threshold}$, such that we are confident that a cracked wire would fall within the *worst-wire proportion*, a one-tailed *t*-distribution at a given level of confidence, $M_{threshold}$ is used as follows:

$$M_{threshold} = \mu + (t_\alpha \, \sigma) \tag{11-3}$$

where μ and σ are, respectively, the mean and standard deviation of ultimate elongation of the set of cracked wire specimens, while t_α is obtained from t-distribution tables at a given level of confidence.

In the following section, condition of any wire is determined as broken, cracked, or intact.

11.4.4 Determination of Wire Condition

The possible outcome of the condition of the wire as broken, cracked or intact is treated as a discrete random variable, X, such that:

$$P(X = x_j) = P_j; \quad j = 0,1,2; \quad \sum_{j=0}^{2} p_j = 1 \tag{11-4}$$

where, x_0, x_1, and x_2 represent in this case broken, cracked and intact wire respectively. The probability of realizing a broken wire is defined as p_0, while p_1 is the probability of realizing a cracked wire, and $p_2 = 1 - p_0 - p_2$ is the probability of having an intact wire. The probability of broken wires, p_0, in each panel is determined based on the number of wires found broken during inspection. On the other hand, the probability of cracked wires, p_1, is determined from assessing the ratio of cracked samples, based on the fractographic evaluation of the fracture surfaces of all wires tested in tension. For determination of the probabilities of broken and cracked wires, p_0, and p_1, in the investigated panel, the effect of broken and cracked wires in the adjacent panels is considered.

When a wire is cracked or fractured in a given panel, it redevelops its load carrying capacity after a certain length, known as the redevelopment or recovery length. While a wire that contains a crack does not lose its entire capacity to carry load, part of its load carrying capacity is lost due to the presence of the crack. The recovery length concept is applied to cracked wires, whereas a cracked wire in a given investigated panel would regain its full load carrying capacity at the end of the recovery length.

The following section introduces a discussion on the recovery length and assumption made in this appendix.

11.4.5 Wire Recovery Length

When a cable wire under tension breaks, or cracks, at one location there is sufficient friction within the cable that at some distance from the location of break, or crack, the wire sustains the same tension as if it were unbroken or uncracked. The frictional forces develop due to the radial pressure applied by the taut wrapping wires and cable bands. Additional pressure and friction are generated near the location of break or crack in wire due to the Poisson's effect. Because the cable is restrained from lateral expansion by the radial pressure provided by the wrapping wire and cable bands, the Poisson's effect increases the inter-wire contact forces. One can postulate a length over which all these frictional forces cumulatively equal the full tensile strength of the wire which had broken or cracked. This length is defined as the clamp or recovery length.

Results of analysis by contact-stress theory showed that the recovery length for an 18" diameter cable with 7,697 parallel wires is 5ft [E.7]. The conclusions of the study demonstrate the effectiveness of wrapping wire and cable bands in redeveloping the strength of a defective wire. It is important, however, to consider the effect of slippage of the wrapping wire or cable bands on the recovery length. Slippage in the cable band may occur due to loss of tension in cable band bolts. This would subsequently reduce the restored load carrying capacity of a broken or cracked wire, at the slipping band. This problem has been encountered on many suspension bridges. In 2009, failure of nine heavy-duty nuts of cable band bolts on a major suspension bridge has been reported. All the cable band nuts and bolts were replaced in the late 1990s as part of a project to replace the suspender ropes. An investigation into the failure of the nuts has identified a number of design and specification decisions and construction methods that may have contributed to the cracking, including replacement of the original bolts with metric versions with a thinner section than the originals. The nuts are small compared with similar ones on other suspension bridges. Another factor was the use of a higher grade of steel, which meant that the nuts are less ductile, i.e., more brittle, than the originals. Misalignment of washers may have led to uneven loading in the nuts. In addition, the protective coating was inadequate and allowed moisture to cause damage.

To account for possible slippage of unwrapped external wires and cable bands, and considering gaps of broken wires that were measured during different cable investigations, a broken or cracked wire is assumed to redevelop its full load carrying capacity after two consecutive panel lengths, at each end of the investigated panel. Therefore the effective number of broken or cracked wires in a panel under evaluation includes wires that are broken or cracked in the investigated panel in addition to the number of broken or cracked wires that are not developed in two flanking panels at each side of the investigated panel.

11.4.6 Broken Wires

The number of broken wires in the exterior ring of the cable in each panel is those identified broken upon the removal of the wrapping wire prior to the wedging operations. Interior broken wires are those uncovered during the wedging operations. As explained earlier, the interior broken wires are typically determined in an eight-wedge pattern, in each panel.

11.4.6.1   Exterior Broken Wires

Wires found broken along the exterior ring of the cable are readily recovered upon the removal of the wrapping wire in each panel.

11.4.6.2 Interior Broken Wires

The total number of inspected interior wires represents a small fraction ($\approx 10\%$) of the wires in the cable cross-section. Therefore in the following analysis, the probability of broken wires, p_0, is assessed based on the observed interior broken wires, in each panel. The proportion of broken wires, p_0, in each investigated panel, is determined by dividing the number of interior broken wires divided by the number of inspected wires.

11.4.7 Cracked Wires

The probability of cracked wires, p_1, in each panel, is determined by dividing the number of cracked wires by the number of tested wires sampled from each panel. As mentioned earlier, cracked wires are identified by microscopic examination of all the wires sampled and tested in each panel.

11.4.8 Strength Evaluation using the BTC Method

In the evaluation of the cable strength in each panel, per the BTC method, the input data consists of the following:
- Probability distributions and correlation for test data for intact wires, $[\varepsilon_e, E, \varepsilon_u, \sigma_u]_{Intact}$
- Probability distributions and correlation for cracked wires, $[\varepsilon_e, E, \varepsilon_u, \sigma_u]_{Cracked}$
- Probability of broken wires, p_0
- Probability of cracked wires, p_1

Using the above input data, the stress strain curve for each wire is constructed. All the wires in the cable cross-section subjected to the same strain. The strain is applied in increments, and the wire fails when it reaches its ultimate strain. Failed wires are discounted from the strength calculations. This process is repeated for the entire set of wires until all the wires reach their ultimate elongation. The load carrying capacity for the cable reaches zero at maximum elongation. The estimated cable strength is the maximum load calculated.

11.5 BTC METHOD FORECAST OF CABLE LIFE

Main cable wires degrade and suffer reduction in load carrying capacity over time. The BTC method forecasts cable degradation as a function of wire mechanical properties, and time.

As explained earlier, the cable cross-section is divided into three groups, namely intact, cracked and broken wires. The broken wires group has no load carrying capacity in the investigated panel. To establish rate of degradation, it is necessary to estimate the time for onset of degradation for both intact and cracked wires, time at which degradation is triggered, and proportions of cracked and broken wires. This is achieved based on data collected under current and previous investigations.

11.5.1 Forecast of Degradation in Intact Wire Strength

Degradation of the strength of intact wires is estimated, based on wire test history, however limited, at different points in time, using the following model:

$$\frac{F_2}{F_1} = f(\kappa, t_1, t_2) \tag{11-5}$$

where F_1 and F_2 are the breaking loads corresponding to the two breaking times t_1 and t_2, and κ is a degradation kinetic that depends on environment, strength and time.

11.5.2 Forecast of Degradation in Cracked Wire Strength

Continuous loading on the cable leads to crack growth in the cracked wires, leading to fracture when the crack depth is such that the stress intensity factor is equal to its critical value, known as the fracture toughness. Due to environmental degradation, the effective fracture toughness is reduced resulting in reduced strength of cracked wires. The environmental degradation is manifested in the appreciable reduction of the strain energy density, W_0, and effective fracture toughness, as first introduced for bridge cable wire in Reference E.8. Therefore fracture toughness and strain energy density measurements provide important information regarding the strength degradation of cracked wires. The strength of the cracked wires at time t_2 is assessed based on the measured effective fracture toughness at time t_2. The strength of the cracked wire proportion at time t_2, $(\sigma_{ult\ cr})_{t_2}$ is then given by:

$$(\sigma_{ult\ cr})_{t_2} = \frac{(K_c)_{t_2}}{Y\left(\frac{a}{D}\right)\sqrt{\pi a_c}} \tag{11-6}$$

where $(K_c)_{t_2}$ is the effective fracture toughness at time t_2, a_c is the critical crack depth and $Y\left(\frac{a}{D}\right)$ is the crack geometry factor.

The estimate for effective fracture toughness of degraded wire, at time t_2, is a function of the strain energy density, which in turn, is evaluated from the stress-strain curve [E.8]. A relationship between the fracture toughness and the strain energy density for a bridge wire was first introduced by the BTC method, as follows:

$$K_c^2 = \beta W_0 \tag{11-7}$$

where β is a function of the elastic properties of the material. The effective fracture toughness, $(K_c)_{t_2}$, at time t_2, is estimated from:

$$(K_c)_{t_2} = f((K_c)_{t_1}, (W_0)_{t_1}, (W_0)_{t_2}) \qquad (11\text{-}8)$$

where $(W_0)_{t_2}$ is the corresponding strain energy density at time t_2, while $(W_0)_{t_1}$ and $(K_c)_{t_1}$ are, respectively, the strain energy density and effective fracture toughness of the wire material at time t_1. Forecast of degraded strengths for intact and cracked wires, as well as updated proportions of broken and cracked wires at different points in time are utilized to build a cable strength degradation curve and estimate the remaining service life of the cable.

11.6 SENSITIVITY ANALYSIS

The input values and assumptions of probabilistic models are subject to uncertainty. This section presents sensitivity analysis for the estimated cable strength due to uncertainty in the inputs.

The purpose of sensitivity analysis is to:
  (i)   identify the key inputs which influence the estimated cable strength.
  (ii) assess whether the estimated cable strength and the decision making process are likely to be affected by such uncertainties.

To conduct sensitivity analysis, key inputs are identified. The values for those inputs are changed above and below a specific base value for the cable strength. The effect of each input is changed at a time, while the other inputs are kept at values corresponding to the base value, and the cable strength corresponding to changed input is then assessed. This process is repeated for the different inputs, above and below the base values, and the effect of uncertainty in each input is quantified. Conclusions are then made about the ranking of sensitivity of cable strength to uncertainties in different inputs.

The following section identifies the key inputs subject to sensitivity analysis.

11.6.1 Key Inputs

The following inputs are identified as influential in the evaluation of the remaining cable strength:
- Effect of adjacent panels
- Proportion of cracked wires
- Ultimate strength of cracked wires
- Proportion of broken wires

As demonstrated earlier in this Appendix, the proportion of broken wires is estimated from field inspection findings of broken wires. The proportion of cracked wires is evaluated based on the presence of preexisting cracks in tested wires. From the analysis of inspection and testing results, it is evident that both proportions demonstrate a range of variation and it is important to define the range over which the sensitivity analysis is performed. The ultimate strength of cracked wires is an input that varies depending on the effective fracture toughness at a given point in time. The effect of these inputs on the estimated cable strength is studied by varying the value of inputs above and below a base value.

The base value for this sensitivity analysis is the expected value of cable strength in the investigated panel, which is assessed based on the effect of two adjacent panels at each end, with proportion of broken wires, p_0, and proportion of wire cracking, p_1. To study the effect of proportion of cracked wires, p_1, a proportion value Δ is added to p_1, while all other inputs are kept at the base value. The following set of inputs will be used to assess corresponding cable strength:

- Number of adjacent panels: 2 panels
- Proportion of broken wires: p_0
- Proportion of cracked wires: $(p_1 + \Delta)$
- Ultimate strength of cracked wires: $(\sigma_{ult\,cr.})$

The cable strength is estimated based on the above set of inputs. Then, the cable strength is estimated once more, but with proportion Δ being subtracted from the proportion of cracked wires, while again the other inputs are kept at their base values, as follows:

- Number of adjacent panels: 2 panels
- Proportion of broken wires: p_0
- Proportion of cracked wires: $(p_1 - \Delta)$
- Ultimate strength of cracked wires: $(\sigma_{ult\,cr.})$

The same process is repeated for all other inputs using the same Δ, above and below the base values. The reason for choosing the same value of proportioning, Δ, is to ensure the validity of conclusions drawn regarding relative sensitivity of assessed cable strengths to uncertainties in different inputs.

11.6.2 Sensitivity Indices

The BTC method defines a sensitivity index as a number which gives information about the relative sensitivity of the estimated cable strength to different inputs of the model. The sensitivity index (SI) is given by Hoffman and Gardner [E.9]:

$$SI = (D_{max} - D_{min}) \qquad (11\text{-}9)$$

where D_{max} is the output result when the input in question is set at its maximum value and D_{min} is the result for the minimum input value.

The calculated sensitivity indices provide ranking for the analyzed inputs. This ranking provides the bridge owner with a clear picture of how the different inputs affect the estimated cable strength by calculating the range of variation in the cable strength corresponding to the range of variation for each input.

11.7 APPENDIX E REFERENCES

E.1. Mahmoud, K. (2009). "Method for Assessment of Cable Strength and Residual Life," United States Patent 7,992,449, August 2, 2009.

E.2. Lichtenstein Consulting Engineers, Inc. (2001). NCHRP Web Document 28: Manual for Condition Evaluation and Load Rating of Highway Bridges Using Load and Resistance Factor Philosophy, NCHRP Project C12-46.

E.3. Hopwood, T. and Havens, J.H., (1984a). Corrosion of Cable Suspension Bridges, Kentucky Transportation Research Program, University of Kentucky Lexington, Kentucky.

E.4. Hopwood, T. and Havens, J.H, (1984b). Introduction to Cable Suspension Bridges, Kentucky Transportation Research Program, University of Kentucky, Lexington, Kentucky.

E.5. Hopwood, T. and Havens, J.H, (1984c). Inspection Prevention, and Remedy of Suspension Bridge Cable Corrosion Problems, Kentucky Transportation Research Program, University of Kentucky, Lexington, Kentucky.

E.6. NIST, (2003). Engineering Statistics Handbook, *U.S. Commerce Department's Technology Administration*.

E.7. Raoof, M. and Huang, Y.P. (1992). Wire Recovery Length in Suspension Bridge Cable, *ASCE Journal of Structural Engineering, Vol. 118, N0. 12*, December, 1992, 3255-3267.

E.8. Mahmoud, K.M. (2003). Degradation of Bridge Cable Wire Due to Stress Corrosion Cracking and Hydrogen Embrittlement, International Mesomechanics Conference, University of Tokyo, Japan.

E.9. Hoffman, F.O. and Gardner, R.H. (1983). Evaluation of uncertainties in environmental radiological assessment models, Radiological Assessments: A Textbook on Environmental Dose Assessment, J.E. Till and H.R. Meyer (editors). US Nuclear Regulatory Commission, Washington D.C., Report no. NUREG/CR-3332, pp. 11.1-11.55.

www.ingramcontent.com/pod-product-compliance
Lightning Source LLC
Chambersburg PA
CBHW082126230426
43671CB00015B/2820